Man Versus Mother Earth

In Loco Parentis

John Aldrick

The natural systems of the earth, its atmosphere and seas,
how man is now affecting them, and the impact this is having
upon himself and the future of mankind

Angel Key
Publications

Man Versus Mother Earth *In Loco Parentis*

The publishers and author cannot accept responsibility for any errors or omissions, however caused, or for loss or damage occasioned to any person as a result of the material in this publication.

Published in Australia for John M. Aldrick by Angel Key Publications 2020.

New edition, revised and updated 2020.

ISBN: 978-0-6488270-9-2 paperback
ISBN: 978-0-6488738-0-8 eBook

Published in Australia by
ANGEL KEY PUBLICATIONS PTY LTD
AUSTRALIA
https://angelkey.com.au

A catalogue record for this book is available from the National Library of Australia

Also by John Aldrick:

Funny Things That Happen
Once I Had A Crocodile
A Taxi Through Phnom Penh

Food, Fatness and Obesity
The Epidemic of Malnutrition

Fantastic Vegetable Garden Soils
Grow Your Own! You can!

Field Use Of 4WD Vehicles
Safe and Certain Outback Travel

Is Recovery From Alcohol Really Worth It?
One Man's Doubts

Acknowledgements

Special thanks are due to Sandy and Ross Towell who have assisted me personally and with the publication of this book.

Dedication

I first came in contact with John Aldrick some years ago when he was in the midst of a personal crisis. Although I was originally there to assist him in this capacity, I soon saw there was so much more to this man and we have since become firm friends.

A learned man of science, John loves nothing more than some educated discussion and debate, to challenge one's thinking. His great love of country Australia is obvious and his travels throughout the world have given him unique glimpses into many other cultures and experiences. It is through these travels, his scientific knowledge and worldly experience that he has been able to put this book together. Never have I read any other book that has pulled together the climate change argument so clearly and concisely.

Working with John to bring the book to life has been a real eye-opener for me, not only because he has not shirked from the frightening aspects of the subject but also because of the way he has logically and scientifically brought the information together into a compelling argument - it just makes sense!

Sandy Towell

About The Author

John Aldrick was born in Orange, New South Wales. He Graduated from Melbourne University with a degree in Agricultural Science, followed by a post-graduate Diploma of Education. After teaching in secondary schools, he worked for State and Territory government departments across Australia and in the CSIRO, and gained a master's degree in Tropical Geomorphology at the University of New England.

His career then took a quantum shift to international consulting, operating freelance with international companies around the under-developed world as a natural resources specialist, including assessment and mapping of natural resource systems, landscape dynamics, use of land for best productivity and least environmental deterioration, and land use planning. This graduated to large scale training programs for staff in these countries. He has written more than 40 technical publications and reports from a dozen different countries as well as Australia.

During this work he witnessed the plight of humanity and the condition of the environment around the world firsthand, and in this book his work history and experience shine through. He spent many years domiciled in Darwin and on the sunshine coast, and now lives in his favourite State of Queensland.

Author's Note

"Mother Earth" was the first Greek God, or Goddess, called Gaia, considered to be a primordial deity. She was seen as a personification of the earth, was associated with the seas, the moon and the stars, and was regarded as the ancestral mother of all life.

The sub-title "In Loco Parentis" is Latin for "in place of a parent" with a positive connotation. In the context of this book "Parentis" means the essential Mother Earth with all the complexity and interconnectedness of her natural systems. "In Loco" implies the alteration or replacement of those natural systems by a new and different parent, Man, with all his deleterious and decidedly negative impacts.

Xavier Herbert, Australian author of the Classic tome "Poor Fellow My Country" and winner of the Miles Franklin Award (1975) wrote; *"The true salvation of men's souls can only come through an all-abiding love for the wondrous thing he owes his origin to – Mother Earth".*

Preface

The greatest plague of all time is upon us. It is not some uninvited affliction such as intractable disease or foul pestilence, it is man himself. It is a simple battle, the inherent makeup of mankind competing against the home that has spawned and nurtured him, the natural biophysical world. Man is undoubtedly a product of the land and its nature, but the tide has turned. Man no longer lives in sustainable consonance with nature, he is now in direct competition with his own supporting environment.

The main driver of this is population growth. World population has trebled in the last century. All animals (and plants) are genetically programmed to reproduce and man is no exception, but that imperative seems to be the last thing that mankind can relinquish. He continues to reproduce unthinkingly, no matter how dire his circumstances may become. Unless there can be a curb upon population growth the human race may bring about its own destruction. In terms of impact, the United Nations has warned that we must stop the continual loss of biodiversity or face our own extinction.

The essential facilitator of this explosive growth has been man's discovery and exhumation of the energy giant, fossil carbon. Fire, since its domestication, in all its forms, has fuelled the advance of mankind, just as the wheel, in all its forms, has facilitated its technology.

Competition between individuals for habitat and resources is a natural means of gaining a reproductive advantage, but man is a very aggressive species and also competes for ethnic, religious, cultural and other essentially tribal reasons. In humans, war goes far beyond competition; man is continuously at war with his own species. War is as basic a part of man as his reproductive imperative, it is innate within his makeup.

The reproductive and aggressive capacities of mankind are destroying his home on earth, and possibly, eventually, himself.

Contents

Central Theme

The Anthropocene

The major problems that now face mankind have developed in a remarkably short time. Today we are officially in the Holocene or Recent geological Epoch, which began 11,700 years ago after the last major ice age. From that time sea levels continued to rise until about 7,000 years ago, and then the climate began to stabilize. The Anthropocene is a proposed new Epoch dating from the commencement of significant human impact on the Earth's ecosystems. It is distinguished as a new period either within or after the Holocene.

It has been argued that human impact upon the earth began approximately 6,000 years ago when conditions on earth became favourable for human occupation, with the development of farming and sedentary cultures. At or soon after this point, humans had dispersed across all the continents except Antarctica. During this time humans developed agriculture and animal husbandry to supplement or replace hunter-gatherer subsistence. These innovations were followed by a wave of extinctions, beginning with large mammals and land birds, driven by both the direct activity of humans such as hunting, and the indirect consequences of land-use change for agriculture. Some consider that much of the environmental change is more recent, a direct consequence of the Industrial Revolution. However, even at that time the overall impact of humans remained relatively small.

It is now clear that human impacts escalated markedly after the second World War, which was the beginning of the nuclear and technological age, about 70 years ago (from 2018). The first war-time atomic bomb was detonated in 1945 and signalled the start of rapid economic growth and a change in the whole fabric of the earth, leading to an unprecedented, abrupt change in climate. During this time humans have manipulated their environment to become more amenable to them, and the population of the world has exploded. In the last 60-100 years there has been a meteoric rise in the number of people on the planet, and they have swarmed across the world like a plague. There is now overwhelming global evidence that during this time atmospheric, terrestrial, marine, hydrological, biosphere, ecological and other earth systems and processes have been significantly altered by human activity. The Anthropocene recognizes that humans are now determining the future direction of the planet.

A geological time interval has traditionally been defined by specifying where it appears in the sequence of rock strata, which can be dated. However, the time period for the Anthropocene is very short, and difficult to pin down to evidence in the geological record. Nevertheless, there are plausible indicators outside this record. Anthropogenic changes in the land such as soil erosion following clearing, the effects of repeated cultivation, the accumulation of fertilizer and pesticide residues, and inclusions of human artefacts can be recognized. Rocks fused with plastic, first discovered on a Hawaiian beach in 2014 could constitute geological evidence. An elevation in background radioactivity, universal contamination and pollution, increased greenhouse gas emissions and the melting of glaciers and the polar icecaps would be other indicators. There is no problem in dating these events, we can remember them.

Mega-problems of the world

Many of the world's people are aware that we are facing major problems and know something of the impact they are having upon us and the earth we live in, but it seems quite difficult to address them, after all, they are mega-problems. What many people don't seem to realize though is the extent to which these problems are global, and how intransigent they really are. These are not just local issues; no matter where they arise, they span the world and affect all its inhabitants. After a time of perfect conditions on earth for humans we are now seeing rapid change. Within the ancient, finely tuned interconnectedness and balance of all biophysical systems on earth man has become a global force such as the world has never known. He is an incredible animal. The essential distinction between man and all other forms of life is that man can think, communicate with precision and detail, and can choose what he will do. He is intelligent, long-lived, and has 20 years of neoteny in which to learn from his parents and peers. He is very adaptable, and has successfully occupied every habitat on earth. There is no top predator able to contain him. Man is the perfect replicator.

Humans are programmed to do two things very well; reproduce, and make war, and the former of these is the one single cause of all the mega-problems of the world, embracing the second. There are already too many people on the planet. Population growth is the main driver of world climate change, resource depletion, global pollution, disease transmission and societal disharmony. Because man is so long-lived there are commonly two or three generations alive simultaneously. People are very proud of the numbers of their children, grandchildren, and great grandchildren, and protect and shelter them assiduously. Queen Elizabeth II herself had 8 grandchildren and 8 great grandchildren; other regals have been similar. Some women are known to have had 20-30 children, and a dozen or so is common. The oldest woman in the world died at the age of 117 years (2018), with 116 consequent descendants. There are reports of much older people in the past.

The median-age demographic which summarises age distribution shows that the median age of the population is much lower in the less economically and socially developed countries, especially the highly populous ones. African countries have the lowest median age, from 14 to 17 years of age, and many of their residents are living on borrowed time. Middle Eastern countries lie between 15 and 18; Afghanistan is 18.8, Yemen is 19.5. South American countries are between 25 and 35. In contrast, most Western countries such as Europe and America have a median age of 38-40 which is more than double that of the lower ones; Japan's median age is 48. The prospect is that a majority of undisciplined, hedonistic youth with all their inexperience, improvidence and immaturity will be eroding cultures that have long recognized the depth of awareness, understanding, knowledge and sagacity of older age, and this may sway the course of human history. Much of south and south-east Asia and the Australian Aborigine are exceptions; Aboriginal "Elders" are revered, looked up to and protected.

World population has increased from 1.6 billion in 1900, 118 years ago, to 2.9 billion in 1958 (an increase of 80%), and in the last 60 years from 1958 to 2019 to 7.7 billion, a further increase of 166% over less than one human lifetime, the Anthropocene. In our living memory, we have made 4.8 billion more people. That's as many as the total 2019 populations of China, India, Africa, the United States and South America put together! In 60 years! In that time, Australia's population has increased from 10 million to over 25 million. Every year, a city the size of Canberra (450,000 in 2018) is added to the Australian population, 180,000 of whom are immigrants or refugees. Roughly

83 million people are added to the world's population each year. Globally, world population is predicted to rise to 9.8 billion by 2050 (UN Intergovernmental Panel on Climate Change Special Report, 2018). Unless this tide of humanity can be arrested, the world will be fiddling while Rome burns.

This is why our cities are becoming increasingly congested. In 2018, Delhi alone had the same population as the whole of Australia, and Shanghai and Beijing were not far behind. Jakarta has 30 million. Greater Tokyo is the world's most populous city with 38 million. Efforts to cope with this population explosion by providing ever more infrastructure such as roads, airports, schools, hospitals and houses are treating the symptoms, not the cause, and further depleting the environment. It is also why governments worldwide continually strive for 'economic growth'.

The overriding problem mankind now faces is the widespread destruction and pollution of terrestrial systems and resources by a massive plague of humans. Some say the cause of the problems of the world is over-consumption of resources, but that is a symptom, consumption is generated by people. The political slogan "Climate change is killing people" fails to recognize that people are the cause of it. Of course, babies are a wonderful and essential part of the human adventure, but it needs to be acknowledged that especially in the underprivileged world much procreative activity is more to do with basic sexual gratification than caring, responsible sentiment. However, it is not a personal issue, it is the continual, rapid, global increase in numbers that is the problem.

This alarming growth and the whole of modern civilization is predicated upon energy released by the unearthing and burning of fossil carbon, which was originally photosynthetic, sequestered by forests during the Carboniferous Period, and this has been the main contributor to global atmospheric pollution and climate change.

Life has existed on earth for 3.5 billion years. Humans are just another kind of life, but the most destructive form of it that the world has ever seen. Over-population is causing not only human deprivation and suffering, but also global warming and sea level rise; increasing super-storms, mega-fires, heat-waves, droughts, cold spells and floods; the over-consumption and destruction of finite land and biosphere resources; pollution of the earth, the seas and the atmosphere; over-fishing of the oceans; extinction of species; the looming disappearance of coral and its magnificent dependent ecosystems; congestion of our cities, roads, ports and seaways; all of which contribute to social unrest and war. Over-crowding of any animal species eventually results in aggressive behaviour and conflict, and it seems that humans are no different.

Humanitarian crises have become so serious that their immediate resolution has begun to overshadow consideration of their causes, which are not only over-population, but tyranny and abandonment by their own despotic governments, now exacerbated by the effects of global warming. World attention is focused on the immediate plight of these hapless people, but consideration of the causes of it is being put aside. The problem is not just that so many people are dying of hunger and malnutrition, it is that humans continue to produce more and more of them; that imperative seems to be the last thing that mankind can relinquish. A feature of every human population living in squalor and deprivation is the preponderance of very young children. There is a never-ending tide of additional babies continuously in the pipeline, to be included in their turn in the register of the starving, with nothing but hopelessness and death before them. Aid organizations trying to help them, and country politics trying to resolve the impacts are being overwhelmed by the scale of the problem, but it will never be resolved without attention to its primary driver, profligate human reproduction.

An increasing humanitarian problem is that of widespread famine and dislocation creating great migrations of whole ethnic and social groups due to political persecution and oppression, or seeking freedom and a safer and better life, sometimes even life itself. There are two worlds; a Western one, composed of cosy, complacent consumers coupled with a commerce that covertly exploits them, and an underprivileged one of deprivation, hunger and hopelessness, overseen by authoritarian and tyrannical governments. The increasingly desperate flights of refugees and illegal immigrants from this latter group has become a massive, worldwide clash between the haves and the have-nots. These people are mostly of African, South American or middle eastern origin, and their movement is always towards Western civilizations such as the UK, Europe, the USA and Australia. Standards of living, safety, and liberty are poles apart in these two worlds, the fugitives know it, and will resort to extremes to successfully transfer. Alongside these mass migrations the lucrative business of people smuggling is flourishing. Resolutions and Pacts by the United Nations and others seek only to recognize the "human rights" of these refugees, which do need to be addressed, but that on its own takes little account of the primary cause, rampant population growth.

There may not be a solution to the overpopulation problem. The human love of and protective attitude towards their babies is absolutely extant. Given the range and disparity of global socio-economics, and the obstacle of relative scales of time, if anything at all can be done, raised standards of living, better education, a more widespread knowledge of birth control, or widespread sterilization, perhaps by cyber implants, might eventually help. Intractable contagious disease, especially in highly populous countries, or martial mega-conflict may intervene as well, possibly insurmountable air and water pollution, or some of the earth's internal processes such as massive eruption of one of the world's super-volcanos. Religion seems to have failed, in fact has exacerbated the problem by espousing the "multiplying" of humans and opposition to birth control. A further problem is that no-one is prepared to cite population pressure as an issue, even if they knew, because that would be politically incorrect, socially abhorrent, and against all religious doctrines.

In some erudite persuasions it has been thought that changing our social attitudes and economic constructs would be the intelligent and civilized way forward, however, this psychology has almost no dissemination; what does have publicity though is commercial sport, celebrity culture and facile politics. It is becoming increasingly apparent that the human genome is that of an animal, and his mind, which is his heart and soul, cannot compete. In outlook, the imperative of biogenetics will triumph over any amount of wisdom, whether we like it or not.

Maybe some sort of alliance will develop between our inbuilt animal instincts and the rapid evolution and spread of new technology. One has to ask, of course, what the driving force behind technology is, and whether any interface with population growth would be benevolent or malignant. Even if plausible new technology for limiting population growth were to be developed, history has shown that there would be overwhelming and likely prohibitive opposition from the church and its adherents. Any such technology would need to be willingly adopted by the people, it could never be deliberately imposed; but it would be virtually impossible to implement because of the thoroughly ingrained tenets of theology, and political correctness.

Many countries are tightening their stance on the ever-increasing illegal entry of refugees. The United States "wall" with Mexico and ban on entry from Muslim countries may have been the start of this. Perhaps what might evolve, awful to contemplate, is that the more civilized, prosperous and aware societies with their knowledge and technology will just jettison the rabble and create two worlds, leaving ignorance and population pressure to eventually decimate itself.

War seems to be an innate part of us. The primal, tribal lore, kill or be killed, is still present at our core as it has been through history. War, somewhere in the world, is always in the offing, and is now way past man's inhumanity to man. Competition between humans starts in childhood, such as picking up teams and setting rules for a clod-fight or a game of marbles. Part of the essence of humanity is territoriality, which engenders cross-border disputes, along with sundry propaganda. Socially, violent political and racial dissident groups agitate and smoulder, and rival gangs, like tribes, claim themselves to be competitors or foes and seek each other out and feel compelled to brawl. Not much different from chimpanzees. We even have rules for war, and we can perpetrate war crimes! We are all "boys" and "girls" socially until it's war, then we are all "men" and "women", which shows the seriousness with which we approach war. It may be that there is an underlying human ethos in combative contact sport; aggressive "sport" may be a placation and substitute for war, as evidenced by the widespread use of gestures and expressions of hostility towards a rival team. In some countries (e.g. Indonesia) sports people openly claim that they would fight or even kill supporters of an opposing team; how tribal is that?

International detente is now significantly more about military might than about trade. Conflict is world-wide, with advanced modern weaponry now in the hands of many nations. Some have become nuclear armed, and if these weapons are used, we won't have to wait for climate change to kill us. Man is making ever larger and more powerful war machines; some nuclear-powered aircraft carriers are 400 metres long, weigh 100,000 tons, travel at 55 kilometres an hour, can stay at sea for 20 years without refuelling, and can fire 40cm diameter nuclear shells. Lethally armed intercontinental ballistic missiles can quickly reach any place on earth. Modern warfare leaves whole major cities that took centuries to build reduced to rubble, and thousands of people displaced or killed. And now there is serious contemplation as to whether war could be conducted in space.

International trade in arms is rapidly increasing. Why do we continue to provide sophisticated deadly weapons to peoples with tribal minds? The result is a new kind of abhorrent; high-tech tribalism; in African cattle-stealing disputes the AK-47 has now replaced the spear. Animals naturally seek an extension of their influence and many do so with loud noises that penetrate for some distance, others by scent marking that can persevere for prolonged periods, but man has radically raised the bar. Guns are an extension of the killing arm of man, and strongly linked to an expression of his ego. The socially, ethnically, and religiously tribal countries are having a ball, like children at Christmas, cavorting around with their new modern toys; tanks, guns, rockets and bombs, not even pausing in their glee to thank their Father Christmas, the trillion-dollar defence (offence?) export industry of the supposedly civilized Western world. The only sanction that will work against rogue warring countries and dissident States is a worldwide embargo upon all sales of arms to them, including the multitudes of war-friendly vehicles. With such an embargo, all war would cease once existing stocks of weaponry became depleted, because those countries cannot manufacture their own sophisticated weapons or vehicles. But the Western world will not do so because they will lose money or access to oil, or by virtue of competing political strategies. To sell arms to a dissident State and then provide military assistance to its protagonist is the height of political hypocrisy. America, Russia and China are in competition for sales of arms and if one of them ceases doing so, one or another of them will step in to fill the gap.

The old Gallic/Saxon ethic of the "Glories of Battle" has now been usurped by indiscriminate, technology-aided mayhem and carnage, civilians being innocent victims. Competition between and within species has been an evolutionary driver of survival, but having survived, what does man do now, how can we turn this thing off? One can hold out little hope for a species at war with itself. As Abraham Lincoln famously said, "A house divided against itself cannot stand".

One of the most fundamental doctrines of the free world is the separation of the Church and the State. Dissent between religious and political interests is rife in some countries, dividing the population. Northern Ireland has been a classic example. However, world-wide, there is a growing integration of religion and politics, particularly in countries with already authoritarian governments. Without a separation of the Church and the State democracy cannot survive, and religion-driven tyranny, anarchy and war become axiomatic. A predominance of religious fervour over discipline and rationality inevitably results in indiscriminate bombing of whole populations or their total abandonment from succour, it overshadows all humanitarian and environmental considerations and leads only to misery, starvation and death. Countries with no overt religion may be better served in this respect than ones with a predominant religious zeal, such as the very great difference between China and Syria.

Politically, increasing regulation and enforcement by public authorities may have become necessary in this over-populated world, but it militates against personal responsibility and restraint, and may prove to be a precursor of tyranny. Repression and poverty have developed in countries that have a one-party authoritarian government, with a tyrant securing control by coup and monopolizing the country's finances, aided by imports of arms from the Western world, and establishing legions of well-paid personal bodyguards and military forces who suppress and fleece the resident nationals to a state of vulnerability and impotence. These tyrants are not statesmen, their drive is self-importance, to dominate and to rule, not to lead. Nations that call themselves a Democratic Peoples' Republic are always despotic. With increasing desperation and inability to cope there is now large-scale migration towards the more developed countries, drug and alcohol abuse, and an increase in criminal activity, as outlets for frustration and to finance these events.

In the Western world politics is not all rosy either. Unfortunately, the world is controlled by politicians, who are demagogues, rather than analogues of Plato's "Philosopher Rulers" (The Republic, circa 380 BC). Despite there being many institutions of knowledge, truth and justice, all that politicians seem to know or care about is politics, i.e. getting elected to power, and all other considerations hold small sway. Governments cannot address the developing crises of the world because their members are characteristically ill-equipped to understand straightforward science and the basic biophysical drivers of change, and are not in a sufficiently competent position to make value judgments. Their reliance upon political subsets such as parliamentary committees, Senate inquiries, and even royal commissions is prejudiced by their filtering of the outcomes and information releases from these bodies to suit their own party's political aspirations. It is a failure of democracy that candidates can be elected to government on largely emotive issues, despite their knowledge base being seriously limited.

Dishonesty in politics may be legendary but it has now become the norm, with politicians commonly preceding a lie with "the reality is", or "the truth of the matter is". Now, with recent permissiveness, politicians are leading an Age of Truly Blatant Lying. Along with false and deceptive publicity and advertising, lying has become a way of life; responsibility and accountability don't seem to matter anymore. The philosophy is, that said loudly enough and for long enough, a lie gains credibility. And if the facts don't support a lie, blame someone else, use threats, or resort to force. Integrity in politics is an oxymoron.

Multinational industries have also become an issue. The food industry, with its components of supermarket outlets and fast foods is one, and the diet industry with its built-in recidivism is another. The food industries have created a worldwide epidemic of obesity and a whole array of associated illnesses, and the diet industry has responded with a plethora of diet programs purporting to counter them. The pharmaceutical industry has quite cynically and avariciously produced "cures", even rather dubious anti-fat drugs. The manufacturers of drugs and medicines not-so-subtly woo doctors to prescribe and even over-prescribe more and more of their products. These are rich and powerful institutions, much more so than the tobacco industry ever was, and with the public as their gullible sheep they manipulate our almost every move. They are essentially selling pleasure, characterized as "neuromarketing", which is driving an epidemic of obesity and chronic disease. The spreading problems of addiction, depression, heart disease, diabetes, dementia, and other so-called "age related" afflictions are the result. The food-based industries create them, then the diet and pharmaceutical ones come along with programs and drugs to alleviate them, with the advertising media as their obedient pawns. These industries, now aided by their own carefully targeted scientific research into human susceptibility, seem to operate as a monstrous cartel.

Food fraud is booming globally. It is now as profitable as drug running, sale of firearms or human trafficking, there is less danger of detection, and the penalties are much less severe. Substitution of inferior or lower-grade products than advertised is very common. Food is adulterated by replacing a food or ingredient in a product with a cheaper alternative. Many of the contaminants such as methanol in vodka and rum or the contamination of infant formula with melamine (in China) or of rice with plastic pellets are unfit for human consumption, if not downright dangerous to health. Dilution of products, such as glucose solution in honey may not be dangerous but is certainly deceptive. Food fraud has been common in such foods as canned and "fresh" fish, olive oil, tea tree oil, fruit juices, spices, tea, coffee, milk, "organic" foods, wine and meat, to cite a few. The fake packaging of these contaminated products can be indistinguishable from the real thing.

Ecological problems, especially the loss of forests, pollution, extinctions, and imported animal and plant invasions are becoming serious. By these means man has destroyed the inherent value and sheer beauty of many once pristine ecosystems. Pollution of the earth's ecosystems including water, soil, the atmosphere and seas is now at increasingly toxic levels. Waterways within the larger cities in the poorer countries are as bad as open sewers, often completely covered bank to bank with floating plastic bags, household rubbish, human excreta and other debris, and the water underneath is dangerously foul. Even the seas are becoming contaminated; a plastic bag has been seen at the bottom of the Mariana Trench, 11 kilometres under the sea. But plastic is just one culprit, a can of Spam has been seen at 1,500 metres deep on the flanks of the Mariana Trench, and a beer can has been sighted at 3,780 metres. Shipping containers that are inadequately lashed on board are often lost from ships, they float for a while and become a shipping hazard, and are sometimes fired upon by the military to sink them. Many now litter the sea floor. Then there are the huge floating garbage patches, now present in all the major oceans.

The atmosphere is already holding unpredictably hazardous gasses. Air pollution in the big cities of many countries is so bad that it triggers a "red alert" and vehicular traffic and industry need to be temporarily curtailed. Beijing may be one, but

Western cities such as London, Berlin, Los Angeles and New Delhi are amongst them. Lack of wind to blow the pollutants away is blamed, but blow it away to where? The World Health Organization reports that globally, half a million children die each year from air pollution. The countryside is also affected. Across India and south east Asia smoke from millions of cooking fires burning wood, grass or dung is emitting huge volumes of carbon particles producing "brown clouds" that are credited with raising atmospheric temperature over locally affected areas and causing respiratory problems and irritations similar to those of smoke inhalation. In these and other countries, women and children spend the greater part of their day foraging over kilometres for ever scarcer sticks of wood or pats of dung to cook with, and to secure water of any quality to sustain immediate life.

Many of the world's major cities, because of the historical need for fresh water are located in low lying areas, largely alluvial plains. Their expansion has rendered some of the best quality agricultural land unusable for agriculture and is causing economic concern and loss of human life due to catastrophic flooding during severe weather events. The world's largest and most dangerous storms, cyclones, typhoons and especially hurricanes are becoming more severe, and their major impact apart from wind is in the flooding of sub-coastal and low-lying areas where there are now so many large cities.

Another issue is that the numbers of native flying insects have been observed to have crashed to an alarming extent. Research in Europe, especially Germany, shows that over the 30 years between 1989 and 2019 insect numbers have reduced by 75%. Apart from the research, this phenomenon is quite apparent, with indicators such as insect hits on car windscreens which were once abundant everywhere now being almost non-existent. The exception, of course, is plague locusts. In a healthy environment, insects provide food for animals further up the food chain and are eaten by other insects, birds, frogs, small reptiles, and some fish. They also undertake most of the pollination of plant flowers and help dispose of animal remains. Loss of pollinating insects such as bees could seriously disrupt man's food supply. This decline has been attributed to the clearing of forests for agriculture and the destruction of fields of native flowers, and particularly the use of pesticides to control insect pests in monoculture agriculture. The situation is probably global. The rich tapestry of abundant life around us is being replaced by a dismal, barren, soulless wasteland, where no birds sing and no frogs croak. Rachel Carson's "Silent Spring" (1962) is well and truly upon us.

And in direct contrast to that is the rising threat of invasive insect species. Voracious tropical termites (Mastotermes darwiniensis) are spreading southwards from their northern Australian habitat. The Fall armyworm was first detected in Australia in Torres Strait and has spread rapidly across to Western Australia and will probably be here to stay. This "worm" is the caterpillar larvae of a moth that has decimated crops around the world and has the potential to cause major economic damage to agricultural industry. In January 2020 a plague of locusts in east Africa was so large and widespread that it contributed to widespread regional famine. One dense swarm of these locusts has been calculated to eat as much subsistence crops as would feed 35,000 people. It was followed by a second, larger swarm. Another serious plague of them hit northern India farmland. Entomologists have claimed that global warming has provided ideal conditions for the locusts to breed. The red fire ant is another dangerously invasive insect pest.

Contagious human diseases are becoming rampant. The recently emerged COVID-19, a zoonotic coronavirus emanating from China is an example. SARS, severe acute respiratory symptom, was another coronavirus (SARS-COV) with similar symptoms and mortality rate. Other viruses include MERS, HIV (human immunodeficiency virus), Ebola, a viral haemorrhagic fever, the human papilloma virus, poliomyelitis, and the Hendra virus that can fatally infect horses and sometimes humans. A common finding with many of them is that they probably originated in animals, often bats, especially fruit bats, through intermediaries such as snakes and monkeys, and possibly after mutating became able to infect humans. No vaccine against a human coronovirus has ever been developed, anywhere in the world. The main problem is that COVID-19 is a respiratory disease that affects the lungs, and because lung tissue is a type of epithelial tissue in contact with the air it behaves somewhat like our outer skin, and an immune response is accordingly subdued.

When contagious infections, even well-known ones like influenza or cholera, take hold in a densely populated community where social hygiene and medical facilities are less than ideal, they become difficult if not impossible to control. Spread is generally through direct contact, droplet transfer, or from contaminated objects such as door handles, supermarket trolleys and escalator rails. Zoonoses, which are diseases that are carried by animals such as pigs, horses, flying foxes and birds and transmissible to man are becoming more prevalent and intractable.

Some of the less developed countries use large amounts of antibiotics in agricultural production, and there is little education on the dangers of resistance inherent in their over-use. Disease-resistant genes in bacteria and other micro-organisms are now widespread in natural systems, particularly water and are fast becoming an environmental pollutant, and there is evidence that these genes can be transferred amongst different bacteria.

Advances in the use of genetically targeted bacteriophages (viruses that attack bacteria) may provide a solution to disease resistance in the future; these viruses will not harm humans and resistance to them cannot develop, any more than the mouse can become resistant to the cat or the rabbit to the fox. The danger though, as with any gene-modified organism is that they may have propensities not fully tested for. What if, for example, a phage designed to attack Staphylococcus aureus (Golden Staph) went berserk and attacked the absolutely essential bacteria in our colons, all one-plus kilograms of them? Nutritional disaster! Or worse, if such a phage happened to attack Rhizobium, the legume root nodule bacteria? Food supplies would plummet worldwide.

Agricultural weeds are fast becoming resistant to herbicides and are as much a threat to the supply of human food as plant disease is to cropping and horticulture, not to mention invasion of the environment generally. Aggressive forest vines are invading tropical forests worldwide.

All these "mega-problems" are unlikely to occur in isolation as one event or another at a time, but two or several of them may well coincide. A combination of events such as severe storms, droughts, floods, fire, drinking water shortages, invasive insect infestations, agricultural plant and animal disease outbreaks, human pandemics, and war for example may all occur at the same time, and the totality of that would be calamitous indeed.

Along with these problems there are socio-economic issues as well. World debt hangs over us all, and average household debt is increasing in many countries. Australia has the second highest household debt in the world, second only to Switzerland (2018 figures). Standards of living are declining across much of the globe. Our current lifestyle is becoming unaffordable for a growing number of people. In areas where war has not yet completely decimated their cities people are facing inexorably rising costs of utilities and essentials such as fuel, electricity, food and water, the failure of transport and other services, and unabating inflation and currency devaluation. It's not just possession of all mod. cons. anymore, it's struggling to maintain the basics, even to have a house to live in. Increasing resort to alcohol, drugs and domestic violence has been a result, even random mass shootings. Many of the population live in a state of debilitating numbness, designed by commerce and subversive governments with help from the media to lull the world's "sheep" into continuing to part with their money. The public are swamped by unconscionable advertising and tendentious claims by greedy money-hungry companies who don't give a damn for public well-being; it has been called an Age of Financial Tyranny.

Issues that are recent and still occurring include such things as inadequate education; high-level corruption; an increasing tendency to spend more on social welfare and defence; ongoing deforestation; the proliferation of cattle; continued adherence to flawed dogmatic beliefs and paradigms; poverty and human suffering; racial prejudice and tension; tides of refugees; aberrant political stances with resultant unrest, starvation and death; radicalization of new terrorist recruits; the increasing resistance of disease organisms to antibiotics; and the loss of personal freedom to government, commerce and technology. On a longer time scale there are also habitat destruction and extinctions of species; the development of "superbugs"; increasing acidity of the oceans and altered marine biodiversity; universal soil and biosphere decline; pollution of the land, the sea and the atmosphere; increasingly severe weather events; melting of the polar icecaps and concomitant sea level rise; and competition for resources with associated societal disharmony.

And in consonance with all of this there is a diminishment of people's sense of security and social cohesion leading to increasing public dissent and unrest, even loss of respect for human life. One of the now obvious hazards of high population densities is that a regimen of individual and social responsibility can revert to that of a mob mentality. Across the world, at rallies, crowds covering tens of hectares packed wall to wall with humans are now common. Mob behaviour has emerged as acceptable, with crowds of abandoned, frenzied people breaking out in mindless eruptions of social unrest. Massive, non-specific "protests" materialize upon any pretext creating mayhem and chaos, fuelled by a deep underlying foreboding and discontent, and sometimes usurped by organized dissident groups and activists pursuing their agenda. In some South American cities, we are starting to see sizable street demonstrations with blatant daylight looting of retail outlets, not so much for greed as for need, on a scale that is little short of anarchy. Individual miscreants and criminals are increasingly being labelled as having a mental illness when the problem is really widespread societal dysfunction. With increasing population pressure and scarcity of resources there has developed a condition of social permissiveness, with money and sex taking over from morals and scruples. There has been a sad decline in clear thinking and aesthetics generally.

With the advent of the technological and digital age our capacity to mould the world to suit our purposes has increased dramatically, and along with it a whole new vocabulary. Due to the speed of technological change many people within just one generation have become out of touch with developing patterns of life and unable to cope with the avalanche of new social and commercial information, and importantly, employment pre-requisites. Many jobs are now being undertaken by remote or by robots having artificial intelligence, and people cannot easily find employment. There is a distinct generation gap, and older people often have little knowledge of the extent to which most aspects of our lives are now controlled digitally. Computers, the internet, satellites, global positioning systems, email and mobile phones are obvious examples, but modern factories, businesses, agriculture, aircraft, ships at sea, machinery, domestic appliances, motor cars, supermarket checkouts and even children's toys are now controlled electronically. Despite the undoubted advantages of this there are new dangers to cyber security in such areas as credit card transactions, theft of personal details, cyber bullying, computer hacking, avoidance of the law, and new directions in international terrorism and espionage, including interference in other country's military pursuits and even election outcomes. All users of the internet, including email, social media and search engines are now identified and labelled by data-harvesting robots for the perpetuation of commercial greed and crime.

Time is an issue. Man is well able to envision possible solutions to the mega-problems of the world but to implement them is another matter. To undo or rectify the destruction of the past is not easy, it will take time, if indeed it can be done at all. Unfortunately, like Kipling's Bandar-log (the monkeys) we humans can only cope with the time span of our own career service or perhaps our lifetime, but the scale of the incredible destruction mankind has wreaked in the same time frame is orders of magnitude more significant. The ones amongst us who can herald time are the historians, but that is retrospective and not able to implement resolution. We do not seem to learn from past mistakes. It doesn't help that history is always written by the winners.

There may be glimmers of hope for man's excesses in over-population, war, and his socio-economic and biophysical environment, with the advent of eco-villages and communities. These are essentially enclaves, which seek to overcome social disharmony and prejudice and to promote tolerance and environmental sustainability. They are becoming widespread, but each unit is quite small and so much out of the scheme of current social and political flow as to be unheralded and un-appreciated. Unfortunately, these small sparks of activity are widely regarded as just another part of the disliked "greenie" syndrome.

Some Characteristics of The Earth

This Section provides background information on the natural processes of the earth; its land masses, atmosphere, and seas, which are the substrate for the impact man is now producing, and is referred to at intervals throughout the following Sections of the book. Some of these phenomena are directly affected by the activities of man and others are completely independent, and operate regardless of surficial eventualities. These are not to be confused, and the distinction is important in assessing the impact of man. However, even phenomena that do not directly influence the mega-problems that we currently face provide compelling contextual information, and if nothing else, they illustrate that the world is a very dynamic place. This Section is somewhat technical, and some readers may wish to skip it or parts of it and proceed directly to the Section 'The Global Impacts of Man'.

Effects of the earth's rotation

The sun rises in the east and sets in the west because the earth rotates about its axis, or spins, at a rate of one revolution every day, which of course is axiomatic. The earth is 40,000 kilometres in circumference at the equator, and this means that people living on the equator are hurtling around at 1,670 kilometres per hour towards the east. Because the circumference of the earth (its latitudinal dimension) reduces towards the poles this velocity reduces also. At very low latitudes people are moving around at much lesser speeds, in fact at zero velocity right at the poles, a person there would be slowly turning on the spot. Most of us, who live distant from the equator, are travelling at closer to 1,000 or 1,200 kilometres per hour. We do not notice this daily circuit except for the passage of night and day because our atmosphere comes with us, and like all things, we are strongly bound by gravity.

Our calendar is not very accurate. We have 365 days in most years, but a day is not exactly 24 hours long; it is 24 hours and 0.98 minutes long, or slightly under a minute longer every day. After four years this extra 0.98 of a minute adds up to 24 hours, so every four years we have to add an extra day to our calendar to "catch up". So we have 366 days in what we call a "leap year", and that extra day is the 29th of February. A leap year can easily be recognized because it is divisible by 4, as in 2020. The phases of the moon recur on a 28 day cycle so that there are 13 "lunar" months in each calendar year. The Lunar or Chinese New Year, which is based on the lunar year rather than the solar year is the biggest celebration in the world, and far exceeds the size of Christmas.

There are other consequences of the spinning of the earth. One is that the resulting turbulence in the atmosphere creates our weather patterns (the Coriolis effect), including the huge circular storms of cyclones, typhoons and hurricanes and the mid-latitude high and low pressure systems. The circular storms move from east to west, and the highs and lows, which are closer to the poles, go the other way. Another effect is that as a consequence of the centrifugal force of the spinning, the oceans tend to be left behind. Water, being a liquid is able to move, and as the earth spins towards the east the seas bank up on their western side. In the Pacific, this amounts to a constant half a metre or so of water height; more if the westwards-blowing trade winds are stronger than usual. Another is that naval guns firing over very long ranges of, say, 20 kilometres must adjust their aim to allow for not only the movement of the ship, like shooting ducks which have to be "led" to allow for their forward motion, but also to allow for the movement of the earth during the time of passage of the shell. Whatever the motion of the target aimed for, it is also moving eastwards due to the earth's spin, but the airborne shell is not, it is travelling too fast and forcefully to travel around with the atmosphere or under the control of gravity. In this case the adjustment would be about 90 metres. And there are other implications, involving such things as the placement of satellites in orbit around a spinning planet.

Flying towards the east is to go in the same direction as the eastwards spin of the earth. It does not alter the distance to be travelled but it does affect the length of a traveller's day. Because of the spin, the progress of the aircraft towards the sun will shorten the length of one's day, and vice versa if one flies towards the west. Hence our body's diurnal clock (circadian rhythm) is disturbed, by having to accommodate different patterns of night and day from what we are used to, and we feel "jet lag". So depending on which direction we travel around the globe (not longitudinally; we don't get jet lag flying north-south) our new day will be shorter or longer than the one we had before. To increase operational efficiency, planes fly towards the west in daytime if they want to extend the day, and towards the east at night if they wish to reduce the hours of darkness. This means that our bodies have less time for our circadian rhythm to adjust when flying east.

The seasons are caused by the axial tilt of the earth, which is 23.5 degrees, so that as the earth travels in its annual orbit around the sun one of the poles will be tilted towards the sun and so experience summer, and the other pole will be tilted away from the sun and have winter. This pattern reverses halfway through each orbit because the tilt of the earth is always in the same direction no matter which side of the sun it is on, and so summer and winter alternate annually between the northern and southern hemispheres. The sun "moves" seasonally between the Tropics of Capricorn and Cancer, back and forth every year. Of course, it is not the sun that moves, it is the earth. The solstices are the days in each year when the sun is directly overhead on the Tropic of Capricorn in the southern hemisphere, or the Tropic of Cancer in the northern hemisphere. The sun always moves right to each of the Tropics, and then reverses its journey. The Tropics of Cancer and Capricorn are each 23.5 degrees of latitude from the equator, which represents the tilt. In a total of 90 degrees of latitude from equator to pole, in each hemisphere, the sun moves through only 23.5 degrees of it (47 degrees in 180 from pole to pole). Much of this has been known since ancient times, partly due to the discoveries of the Greek astronomer Hipparchus around 130 BC.

Day length changes with the seasons and is determined by whether the sun is moving towards you or away from you. For example, south of the Tropic of Capricorn the days continue to get longer as the sun moves southwards (towards you) until it is directly over this Tropic, which is the longest day, and then as the sun turns and goes back north again (away from you) the days begin to get shorter, and the shortest day occurs when it reaches the Tropic of Cancer. However, between the two Tropics such as at Cairns, which is north of Capricorn, there are two days in each year when the sun is directly overhead. One occurs when the sun passes overhead at Cairns on its way south towards Capricorn, which is one of Cairns' longest days. From then the days get shorter (at Cairns) because the sun is moving away, southwards, and when the sun reaches Capricorn, Cairns has one of its shortest days. As the sun returns from Capricorn and moves back towards the north, the days get longer again (at Cairns), and when the sun is directly overhead for the second time at Cairns it has another longest day. After that, as the sun keeps on moving north, the days keep on getting shorter until it reaches the Tropic of Cancer, when Cairns has its second shortest day. Effectively, any area between the two Tropics has two shortest and two longest days in every year. Monsoonal countries in this zone experience two wet seasons every year, a "little wet" and a "big wet".

Because the earth's orbit around the sun is elliptical, it has points where it is furthest from or closest to the sun, called an apogee and perigee of orbit. On earth, these points do not occur exactly at the solstices. The solstices are the points when the tropics of Cancer or Capricorn are furthest from or closest to the sun, not the poles, so planet earth still has about two weeks to go beyond the solstices before it is right at its apogee or perigee, and then the (theoretically) coldest or hottest times of year occur. These points are known as aphelion and perihelion, which come from ancient Greek, where 'apo' means far and 'peri' means close. 'Helios' is the sun.

The seasonal migration of the sun across the earth results in the zones poleward from the Arctic and Antarctic Circles experiencing six months every year when the sun is either continuously above or continuously below the horizon. These are known as the polar day and the polar night. The process does not happen suddenly but progressively, as the earth slowly orbits the sun. In the northern hemisphere, it begins at the Equinox (in September) when the sun is directly overhead at the equator and moving southwards. At that time, the North Pole has its last sunrise. As the sun continues southwards, the polar night (the area with no visible sun) slowly extends from the Pole and reaches its maximum extent at the Arctic Circle when the sun reaches the Tropic of Capricorn, which is the northern hemisphere winter solstice (in December). Then, as the sun travels back towards the north the polar night retreats, until at the next equinox (in March) it has all returned to normal sunrise and sunset. Over the second six months of the year, as the sun continues to move back towards the north, the northern polar region has a polar day and the southern one has a polar night.

During a polar day the sun is completely visible for 24 hours every day, it never sets at all (the land of the midnight sun). However, although in a polar night the sun is continuously out of sight there is not continuous complete darkness, due to the effects of twilight. Twilight is due to illumination of the atmosphere by a sun that is below the horizon. It begins at the last sighting of the sun and decreases until the sun is 18 degrees below the horizon, which takes a total of about a month and a half. So given two twilights, one going and one coming, the time of total complete darkness is really only about three months.

The earth and the moon interact. Earth's gravity, of course, keeps the moon in orbit. There are two obvious effects of the moon upon the earth. One is that it shines, illuminating the earth at night. The moon does not produce light of its own, it reflects the light of the sun as "moonlight". When the earth's shadow covers part of the moon this reflection is reduced, so there are "phases" of the moon. The other effect is that because it has a large mass the moon exerts a gravitational pull, and on earth this is reflected in the tides. The gravitational pull of the moon creates two tidal bulges in the seas; one occurs directly under the moon and the other on the opposite side of the earth. This second bulge is counter-intuitive; it is a compensatory bulge due to "inertia counterbalance" which is the normal response of a spinning object to maintain its dynamic equilibrium and not wobble. So there are two high tides at the same time, the compensatory bulge one being slightly smaller, and any spot on the earth experiences them both.

The sun also has a gravitational pull which is why the earth remains in its orbit, and it too effects the tides, but its tidal effect is smaller. The moon orbits the earth on a lunar-monthly basis and its tides vary through the month, but those of the sun vary only on an annual basis, because the earth orbits the sun only once a year. The highest tides occur when the sun and the moon are in line and pulling together, called spring or king tides. The smallest tidal range is when the moon's pull is at right angles to the sun's and their combined gravitational pulls are at a minimum, called neap tides. Spring and neap tides occur twice in a lunar month. The highest tides in the world are in the Bay of Fundy, Nova Scotia, which rise to 16-18 metres and when low can retreat five kilometres offshore, and at Anchorage, Alaska, which can reach 13 metres. In a seasonally high tide in the Bay of Fundy more water flows into the Bay than all of the world's rivers combined. Northern Australia has a range of about 9 metres.

Tidal extremes can affect the weather because large rises and falls in sea level displace a lot of air, and this affects atmospheric conditions. This lends credence to the bushman's lore, that there will be rain at the next full moon. These gravitational forces also affect the phreatic water table on land, and springs are known to flow more strongly when the force is at its greatest. There is also said to be an effect upon human moods and sleep patterns, and why not, the human body is 70% water.

Time zones around the world are a construct of humans, because time is of course continuous. The 24 hours it takes for one revolution of the earth have been divided for convenience into segments of one hour so that there are 24 time zones. Each hour represents a span of 15 degrees of longitude, 24 times 15 being 360 degrees. Historically, Greenwich has been used as the reference point or prime meridian. However, for political and convenience reasons the zones are not always strictly adhered to and are moved to suit different country borders, and sometimes split in half. Given that the circumference of the earth at the equator is 40,000 kilometres, each 15 degree division there spans 1,670 kilometres, but near the poles the lines of longitude converge and the distance between each hourly division becomes progressively less, and could be stepped over at each of the poles.

The total length of a day never changes, but the ratio of hours of daylight to hours of darkness does. Working hours are basically in accordance with the time zones and clocks are set accordingly. As the sun moves towards the equator, from either solstice, and summer approaches, hours of daylight become longer and hours of darkness correspondingly shorter. This has led to much debate in countries that extend along an appreciable length of longitude such as Australia and North America because in these countries the time zones have varying amounts of daylight along their length. In summer, the sun rises earlier and sets later, and hours of daylight are longer. This has led to the instigation of "summer-time", where countries advance their clocks by an hour during summer, rather than flexing naturally with the seasons. This has a habit of dividing a country socially in terms of the starting and finishing times of employment. There is a basically sybaritic reason for the adoption of "summer time", because if workers begin their employment a little earlier (by sunrise time) they can stop work an hour earlier also, and so by transferring a morning hour to the afternoon, they have an "extra" hour of leisure after work.

The extinction of the dinosaurs

The dinosaurs were killed by an asteroid that hit the earth over 66 million years ago during the Cretaceous Period. At that time the earth was much warmer, there were no polar icecaps, and the Arctic and Antarctica were covered in thick forest. Dinosaurs roamed wherever there was land, which at that time included the polar regions. They had been in existence for about 20 million years before the asteroid struck. During that time Africa and South America had separated from Antarctica, followed by India and Australia, and all of them were slowly drifting northwards. The asteroid landed just offshore in a shallow sea at a point which is now just to the north of southern Mexico, the Yucatan Peninsula, in the Gulf of Campeche. The crater has now been all but obscured by about 600 metres of pulverized, fractured rock rubble which has been deposited since the impact.

The asteroid was 10-15 km wide and came in at an estimated 72,000 kilometres per hour. It made a hole in the ocean more than 200 km across, and a crater in the ground beneath it 100 km wide and 30 km deep. The impact was like a huge nuclear explosion, equivalent to about ten billion Hiroshimas. The sea was vaporized for 200 km, and boiled as it came back in. As it returned to fill the hole it caused enormous tsunamis, twice as high as today's tallest buildings. To put this into perspective, in the very early geological history of the earth (about 4.5 billion years ago), a space body the size of Mars struck the earth a glancing blow, and the debris it threw out eventually coalesced to become our moon. This impact was so gigantic that it caused the earth to tilt on its axis at 23.5 degrees relative to its orbit around the sun, called obliquity. This tilt now causes the seasons.

When the Cretaceous asteroid struck it created shock waves that spanned the earth. It created a vast radiation fireball of around 10,000 degrees Celsius which raised the temperature globally. All living things within thousands of kilometres were killed within seconds, and 75% of life on earth eventually went extinct. Everything caught fire, and the fire spread around the planet in only a few hours. It was an apocalypse, Armageddon. After only about ten minutes the sky darkened, and in a couple of hours a hot blanket of dust and rock particulates formed around the earth. As this blanket fell back to earth it formed a tell-tale deposit or "signature" around the world that can be recognized today as a yellowish ash a few centimetres thick. This deposit isolated any remaining dinosaurs from the natural soil, and by causing nutritional dysfunction probably contributed to their demise. Then, from being so hot it became very cold. Sulphurous gasses had been released from rocks deep underground, and these, with the dust, caused the sun's rays to be blocked. Temperatures plunged, and it became like a nuclear winter. The earth became like a moonless night, and there could be no photosynthesis. Life hung by a thread; even animals that were adapted to the cold had difficulty. The animals that survived were mostly those able to find cover in sheltered caves on land or in the sea, and wait out food supplies and breathable air. Animals that could fly were better off, if they flew towards the equator. Archaeopteryx was one of the first birds, still a dinosaur, it had teeth in its "beak", but it had feathers and could fly. The larger dinosaurs couldn't adapt or fly, and they all perished.

After the maelstrom, a few dinosaurs including crocodiles and turtles were able to survive because they had the shelter of the water. Frogs, and cockroaches (of course) also survived. With the massive widespread extinctions there was plenty of food for them in the initial period, and they could last for long periods without food anyway. Some of the tiniest dinosaurs (raptors) did make it through and some of these became our modern-day birds, now much changed from their ancestors. Others of today's organisms may have an ancestry many millions of years old, as for example some reptiles (Komodo Dragons, crocodiles), insects (dragonflies, blowflies), and the descendants of ancient monster fish. There are still abundant living marine algal bioherms and bacterial stromatolites virtually unchanged from millions of years ago, which pre-date corals. The calcium "skeletons" of these marine structures can be found in the fossil record as limestone deposits similar to those of extinct coral reefs. Cycads (Zamia palms) which still grow today on old land surfaces, and the Ginkgo tree are also ancient. By comparison, the rice paddies of the Himalayas which are amongst the oldest human structures on earth are only a few thousand years old.

Then there began an age of mammals, and many of them grew very large. The Blue whale, a mammal, is the largest animal ever to have inhabited the earth and weighs in at around 200 tons.

Gravity and magnetic fields

There are two "force field" properties of objects on earth and in the wider universe, gravity and magnetism, and these are properties of matter, or mass. These properties have no bearing on superficial changes in the land or the atmosphere that have been wrought by man. It could well be asked, what is gravity? We don't fully understand what gravity is, but we do know how it behaves. Gravity is a property of any mass; the mass does not have to be ferrous or made of iron, which is a requirement of magnetism. It is an invisible, universal force of attraction. The motion of falling apples and circling planets is controlled by gravity. Mass is the only thing that can possess gravity. Gravity causes mass to have weight. There is a difference between mass and weight. Mass is the quantity of matter in an object and always stays the same, but weight is a measure of the force of gravity upon that matter, and this can vary. A smaller mass has a weaker force of gravity, and objects upon it will weigh less. Due to their different mass, the moon has 16% of earth's gravity, Mars has 38%, and Jupiter, which is larger than all the other planets combined, has 2.5 times the gravity of the earth.

Mass attracts other mass by gravity, but curiously, no mass will repel other mass, in contrast to magnetism. Gravity is geocentric, it pulls towards the centre of the object, it has no positive and negative poles for the force to flow between, as magnetism does. Perhaps counter-intuitively, a feather and a lump of metal released in a complete vacuum where there is gravity will fall at exactly the same rate; on earth, it is the atmosphere that slows the feather down. Astronauts on the moon have the same mass they had on earth, but they weigh much less. The functioning of our muscles, heart, blood vessels and balance system all function in accordance with earth's gravity, and after a sojourn in space astronauts must allow time for their systems to return to normal.

Gravity is a very long-range force; it spans the universe. It slowly diminishes with distance, and at long distances across the cosmos it mainly affects large objects like stars and planets. It is the most powerful, all-pervading force in the universe. The granddaddies of all possessors of gravity, black holes, have "infinite" mass. A black hole has the most stupendous gravity imaginable, and an atom the least. Galaxies, such as our Milky Way, which are immense and numerous beyond comprehension, have inside them one or more black holes. Each is very tiny but has colossal mass and an unbelievable power of gravity, gravity so powerful it can even pull in light. Because they emit no light, they are invisible even to the technologically aided eye; their presence can only be inferred by mathematics, which is the language of science. The stars in galaxies including our own Milky Way are held in orbit by and revolve around their black hole, just as planets revolve around their sun and moons around their planet, all controlled by gravity.

The other "force field" is magnetism. Magnetism occurs mostly in ferrous metals but not only those, a mineral called magnetite has it too, and to a degree so has iron ore. Magnetism is dependent on how the electrons in a material are aligned. Electrons are like mini magnets. Generally, the electrons in (say) iron are arranged randomly, but they can lower their energy state by aligning themselves so that they pull together and repel together, thus creating a composite magnetic force. The more complete the alignment the stronger the magnetic force. Ferrous metal items can be magnetized by another magnet to a greater or lesser extent, magnetism can even be induced in a ferrous metal by an electric current to form an electro-magnet, widely used in industry, so the thing is transferable, like a disease. A compass is a magnet, that if free to rotate will align itself with the earth's magnetic field and enable us to tell north from south. Some iron-rich land masses will affect the accuracy of a compass and cause it to indicate "magnetic" north rather than true geodesic north. The magnetic field of the earth emanates from one pole and enters into the other, like a common metal magnet. However, here on earth things are not all whisked away to the poles to stick there and cling to each other as iron filings do with a common iron magnet, because at any point magnetism is a very weak force; gravity is much stronger and easily holds things in place.

Not all planets have a magnetic field. Earth's moon not only has no atmosphere but no magnetic field. The earth is the densest of the planets in our solar system and its magnetic field is due to the iron composition of its molten core. Jupiter, the biggest planet in our solar system has a mass more than double that of all the other planets combined and a large and enormously powerful magnetic field, about ten times as strong as earth's, thought to be generated by a core of "metallic" or "atomic" hydrogen. The magnetic field of the earth is known to have changed polarity on occasion throughout geological time, so that north has become south and vice versa. Magnetic fields protect a planet from charged particles that stream out from a sun or star by deflecting them outwards along the lines of the magnetic field and preventing them from affecting living things on the surface of the planet. On earth, the interaction of these solar particles with the magnetic field produces the phenomenon of auroras – Australis at the south pole and Borealis in the north.

Interestingly, from space, the sun is actually white, it appears yellow to us because of the effect of the earth's atmosphere.

There is a point near Alice Springs where lines of the earth's magnetic field intersect, I don't know why, but Australia is a very big land mass and perhaps it distorts the flow of magnetism between the poles. It is marked by a datum point on the ground. I have flown directly over it and all gravimetric compasses go berserk, they spin around wildly for about 500 metres and aircraft must maintain their original course until they are through it.

The specific heat of water

Water in all its forms (solid, liquid and vapour) is a major component of the land, the atmosphere, and of course the seas. It is necessary for all life and is a major constituent of the atmospheric "blanket" which causes the natural greenhouse effect that keeps the earth warm. The activities of man significantly affect the role of water in all its states, and this includes impacts upon weather and climate, storms, rainfall, flooding, drought and agriculture.

Temperature and heat are not the same. The actual temperature of a substance and the amount of heat being added to it or taken from it are different. They are related by "specific heat". Specific heat is the amount of heat required to raise the temperature of a unit of mass of a substance by one degree Celsius and is a constant for any particular material. The definition specifies a mass of one gram of the substance, and measures the heat required in calories. Heat is stored in the motion of the atoms or molecules of the substance; at higher temperatures these particles shoot about or vibrate more quickly. Different materials require different amounts of heat to raise the temperature of a gram of them by one degree. Specific heat is not to be confused with latent heat, which is the heat required to change the physical state (solid, liquid, gas) of a material and may be exothermic or endothermic.

Water is the substance that requires the most heat to change its temperature and has been allotted a specific heat of one, and is the standard against which other materials are assessed. But the devil is in the detail. For this assessment, the water must be pure. "Pure" means water with nothing else in it; i.e. no dissolved or suspended salts, gasses, dust, pollen, or other liquids mixed in with it. Metals such as in cars or crowbars have a very low specific heat. Copper needs less than 10% of the amount of heat that water does to raise its temperature by one degree. The same applies in the reverse direction; copper will become one degree colder after losing only 10% of the heat water would have to lose.

Substances made of quartz such as the sand in deserts also have a very low specific heat and can change temperature radically with a small amount of heat gain or loss, which is why deserts are very hot during the day but very cold at night. Swamps are the opposite, because of their high content of water. Coastal areas also remain at a relatively constant temperature due to the moderating effect of the ocean. The atmosphere is also affected. Dry air will change temperature through a large diurnal range, but humid air contains a lot of water vapour and has a smaller difference between daily maximum and minimum temperatures. This is the reason for the large range in daily temperatures from minimum to maximum in inland locations compared with coastal environments.

Rate of temperature change is different also; dry air will change its temperature quickly because it contains less heat, whereas air with a lot of moisture in it will undergo temperature change more slowly. Colour also influences temperature.

Everyone knows that a black car will absorb heat more quickly than a white one and will also radiate heat off more quickly. With persistent application of heat such as with a black car in hot sunshine the car will also absorb more and more heat and so its temperature will rise a lot more than a white one, partly because white is also a better reflector.

So, what would the temperature of this hypothetical gram of water actually be in any given circumstance? That question is not really applicable; and the answer is "it depends"! It depends on all those variables mentioned above. The concept of specific heat does not relate to a substance at any particular temperature, it is purely the amount of heat required to raise or lower the existing temperature of a gram of it by one degree Celsius. But the world of thermodynamics is complex and not for the faint hearted.

Cyclones, typhoons and hurricanes

These are large and powerful weather phenomena that occur as very low-pressure systems along either side of the equator, and as they form, they develop an "eye" around which they revolve. Their central pressure varies from 980 millibars to as low as 880 millibars (one atmosphere is 1,000 millibars; high pressure systems may go above 1040 millibars). The zone of destructive winds in very large hurricanes can be a thousand kilometres across, and the eye itself may be 50 or even 100 kilometres across, more is on record. They develop tremendous energy, and the warmer the seas they pass over the more intense they become. A large hurricane is by far the most powerful event on earth, even a small one releases the equivalent of a ten megaton nuclear bomb every 20 minutes (Hiroshima was only 0.015 megatons). There is enough total energy in one of them to run the electricity needs of the USA for years, if a way could be found to tap into it. Birds caught up in these storms drown in mid-flight as their nostrils are on the top of their beak, or get trapped inside the core and transported for hundreds of kilometres.

These intense storms are driven by the spinning of the earth, its diameter being much greater at the equator than at the poles, causing a swirling effect in the atmosphere. Along with polar vortexes and the mid-latitude high/low pressure systems they are a process whereby heat is transferred from equatorial regions, which are subjected to intense tropical radiation, to the polar regions, which otherwise would become extremely cold.

All these huge revolving disturbances form between the tropics of Cancer and Capricorn, but never actually on the equator. All three types of storm travel from east to west around the tropics, interrupted at times by collisions with continental land masses. Those that form in the southern hemisphere (cyclones) turn in a clockwise direction, and those in the northern hemisphere (typhoons, hurricanes) turn anticlockwise, otherwise they are the same thing. Wind speed is greater if it is additional to the forward motion of the storm. If one experiences both "halves" of the storm the southern half (southern hemisphere) or the northern half (northern hemisphere) has greater wind speed for this reason. If they move outside the tropics they soon lose their eye and much of their intensity, and are downgraded to tropical storms, then tropical lows or rain depressions. These degrading weather systems don't have destructive winds any longer but can produce prodigious amounts of rain. South of the Tropic of Capricorn the nearest weather events to cyclones are what in Australia are called "severe weather events", or if offshore, "east coast lows", which do not develop an eye. They may have wind gusts, as distinct from sustained wind speeds, in the cyclone categories one or even two range but they are not cyclones (or tornadoes!).

Hurricanes are the largest tropical storms. They form in the North Atlantic Sea and pass between North and South America through the Gulf of Mexico and the Caribbean islands into the North Pacific, leaving a trail of destruction in their path. The Americans named them hurricanes before they were known to be circular systems because of the very strong winds that occur in them. Extensive flooding of low-lying areas is also a feature of these storms, and nests of tornadoes are commonly generated in association with them.

A common phenomenon when these storms are in near-coastal situations is rapid intensification. Cyclone Tracy which hit Darwin in 1974 underwent "explosive deepening" or a rapid drop in central pressure when its eye entered Darwin harbour, due to the warmer temperature of the shallow water there. It is not clear what category that one was because the official anemometer was damaged by flying debris during the first half registering a wind speed in the category 4 range; but it may have been greater.

All these storms can have accompanying "storm surges", which are rises in sea level as the storms come ashore. The very strong winds in these systems blow sea water up into a hump that rises above normal tidal level. In addition, the lower the central pressure the more sea level can rise in accordance, and the higher the surge can be. Surges are also higher when the wind is blowing in the direction of movement of the storm, in Australia this is towards the west. A significant tidal surge can be nine metres or more in height which is enough to cover a house, and four metres is common. During tidal surges sea water can invade and flood low lying areas of the coast and may penetrate for kilometres inland.

An interesting thing about wind is the production of raised dust. The dustiest place on earth is the land surrounding Lake Chad in the south of the Sahara Desert, an area known as the Bodélé Depression. The soil here is composed of diatomite, which are deposits derived from ancient plankton. These deposits form naturally on the sea floor where they may be around a kilometre thick. If they are oceanic deposits, the Bodélé Depression was probably once an ancient sea floor. Lake Chad is quite shallow and varies in size enormously, it has filled and shrunk many times. Sometimes it occupies a huge area of more than 25,000 square kilometres but then it may contract to almost nothing and the surrounding land dries out, allowing the prevailing winds to regularly blow up huge dust storms. This dust traverses the world, and much of it crosses the Atlantic to the rain forests of South America. Over the Amazon, this fertile dust is caught up in atmospheric moisture and falls to earth in rain. Because of its origin in foraminiferi, krill and other plankton, it is very rich in plant nutrients. The Amazon rainforest receives millions of tons of mineral-rich African dust every year, which is partly why it is so lush and prolific.

Archimedes, clouds and lightning

We are used to things floating on a liquid where there is a clearly observable boundary between the liquid and the air. Things may float upon the surface of any liquid that can support the weight of the object. Archimedes propounded a well-known Principle; that any floating object will displace an amount of water (or other fluid) equal to the weight of the floating object. One litre of fresh water weighs one kilogram, so if a floating object displaces ten litres of fresh water it will weigh ten kilograms. This Principle can be used to "weigh" very large ships, and the depth at which they float is marked onto their hull as their "Plimsoll Line". Salt water can be used by simply substituting a density value for salt water instead of fresh. Ships made of iron may seem too heavy to float but it is the volume of air inside them that keeps them afloat and air has very little weight, the iron is only a skin, like a piece of wood enclosed in metal foil.

Sea water contains dissolved salt and is consequently denser than fresh water. The more salt that water contains the greater its density and the more a litre of it weighs, and the less of it a floating object will displace. The Dead Sea, on the border of Jordan and Israel is completely land-locked and is 400 metres below sea level; the lowest point on earth. It is ten times as salty as the sea and hence denser, and objects immersed in it including humans float more out of the water. And, incidentally, it is right on a crustal plate boundary and is slowly soaking away. There are others, such as Laguna Cejar in Chili. Another is Mono Lake in Mono County, California which is a saline-soda lake, and the lack of an outlet causes high levels of salts to accumulate in it; it is twice as salty as the sea. Lake Natron in Tanzania is similar, but the solute is not salt, it is carbonates, so it is ill-named; "natrium" is another name for sodium, hence the chemical symbol Na for sodium. This lake has been referred to as a caustic alkaline brine. The Sargasso Sea, famous for its content of seaweed, is completely surrounded by North Atlantic sea water and is only slightly saltier than normal sea water. Owing to slow circular currents, it has not only seaweed, it has now accumulated a high concentration of non-biodegradable plastic waste, known as the huge North Atlantic Garbage Patch.

Archimedes Principle still applies even if the object is below the surface of the liquid. If an object is immersed in water almost but not quite afloat so that it displaces all its own weight, it will not sink but just stay where it is. It would only sink if it was heavier than the weight of water it displaced. Water is essentially incompressible and maintains a constant weight per volume, so even if the object were to be transported down to depth it would similarly stay where it was placed. However, if the object were to be compressible, although its weight would not change, it would become smaller due to the increase in water pressure with depth and would then displace a weight of water less than its own weight and would sink, increasingly faster if the object continued to compress.

Human bodies are over 70% water and that part of them is incompressible, but at depth they compress to an extent because the lungs and other body cavities are crushed to virtually nothing and the body becomes smaller. It would weigh exactly the same but being smaller would displace less water and would sink. A submerged body will eventually undergo the processes of decay and associated production of gas, and in shallow water this may cause it to rise and re-emerge at the surface, but at deeper levels water pressure would compress the gas produced and the body would continue to sink.

Air does not have a clearly observable upper boundary like water, but it increases in density from its outer limits where it has very little density right down to sea level. Air pressure increases with the weight of the air column above it, i.e. with depth, and at sea level air exerts a pressure of 14.7 pounds per square inch or just over one kilogram per square centimetre of surface area. Imagine that; if you look at one square centimetre of your body there is a kilogram of air pressing against it! No wonder planes can fly. Air pressure is referred to as atmospheric or barometric pressure, one "bar" being equal to one atmosphere. By Archimedes Principle, an object placed in the air will sink until the volume of air it displaces has a weight equal to that of the object. Clearly an object must have a very low weight to float anywhere in the air column, and clouds are in that category.

Clouds form because of the moisture content and temperature of the air. The temperature of the atmosphere reduces fairly uniformly with altitude, so that the higher the altitude the colder it becomes. Water vapour is a major constituent of the atmosphere. Warm air can hold a lot of water as invisible vapour, and we record this as "humidity". At a certain height the air temperature drops to where it is too cold to support the water as vapour and it condenses into liquid in the form of very fine droplets, finer than in fog or mist, forming clouds. There is a very clear line where this occurs known as the "dewpoint", sometimes referred to as a 'cap' or temperature inversion. At very high levels in the atmosphere the temperature in clouds is so low that the tiny particles of liquid water freeze to very fine crystals of ice, as in cirrus clouds, often described as mare's tails.

Cloud formation is driven by convection currents or thermals that establish during the day as the sun warms the earth, causing plumes of moist air from below to break through the dewpoint temperature line. Thermals are quite strong; they easily support gliders. In thermals, the upper parts of the clouds billow up and expand into large storm clouds, a condition known as atmospheric instability, a precursor of rain. This is why many clouds, especially cumulo-nimbus storm clouds have a flat bottom and a billowing fluffy top. Clouds can also be formed by "orographic" effects, where moving moist air rises over high hills or mountains into air which is cooler than the dew point. From the air, land masses can be identified by the orographic plumes of cloud that rise above them, but not over the seas.

It is curious that clouds, which are made of water, stay up there. The water in clouds has condensed into a liquid form, but the particles are very tiny. They are so small that the attraction of gravity upon each of them is negligible, similar to the miniscule effect gravity has upon tiny particles of dust or pollen that hover in the air. However, collectively, the water in clouds can add up to quite a lot, and as the clouds get larger, the particles begin to coalesce to form bigger drops, that fall out of the sky as rain. If clouds reach a cold enough temperature, as occurs when updrafts raise the air to higher levels, hail, which is frozen rain, can result during storms. The water in an average cumulus cloud weighs about 500 tons, or as much as a large passenger plane or 100 elephants. Rain clouds are much bigger, and a normal rainfall of 25mm in a thunderstorm would weigh 250 tons over a hectare, and such a storm passes over many hectares. When clouds are moving through the air they have a "flying" effect similar the that of aircraft, their speed is much slower but so is the mass to be supported. It only takes a bit of wind to fly a kite.

In some conditions a very high atmospheric content of water vapour can produce clearly seen and felt rain, from a clear blue sky. This condensation is not due to temperature, but because the amount of water vapour that the atmosphere can contain at that temperature has been exceeded; the humidity has risen to over 100%, and some of it condenses into liquid. Conversely, falling rain can evaporate if it encounters less-humid air before it hits the ground, and in drought conditions this can often be seen happening. At the other end of the temperature scale in below-freezing air temperatures, moisture in the atmosphere freezes to produce tiny particles of ice known as "diamond dust", so that the air is rendered very dry. All these phenomena involve changes in the physical state of water between vapour, liquid and solid. Scientifically, these changes of state release or fix large amounts of heat, known as latent heat, which has an enormous effect upon the weather in various parts of the world.

Fog, dew and frost at ground level are completely different from clouds, they are caused by the temperature of the ground rather than that of the atmosphere. After sunset the land radiates heat until dewpoint is reached at ground level and it becomes cold enough for atmospheric moisture there to condense into fog or dew, and if it becomes even colder, to frost. The longer the time available for radiation to occur the more severe the frost. Cold air, being denser, tends to drift to lower-lying areas, making frosts there more severe. Then, with the arrival of the sun, dew, fog and frost evaporate and conditions become more conducive to human comfort and plant growth.

Interestingly, storm clouds contain very small particles of matter, and the heavier of these, logically, accumulate at the base of the cloud, and adopt a negative charge, whilst the lighter ones rise to the upper areas and adopt a positive charge. This potential difference is responsible for static electrical discharges of "lightning" within the clouds. The negatively charged particles at the base of the cloud are also responsible for lightning strikes down to the ground, which has a positive charge. Electricity (consisting of electrons) is transmitted via the shaft of lightning. Most ground strikes occur around the perimeter of thunderstorms, so we experience them mainly at their leading and trailing edges. In the middle part of the storm, where most of the rain is, lightning tends to be up within the cloud mass. In a lightning strike, the shaft of current is only about as thick as one's thumb, but it ionizes the air around it so that it appears much thicker. Lightning produces nitrates ($NO3$) from atmospheric nitrogen (N) and oxygen (O), which dissolve in rainwater and fall to earth, where they act as a plant fertilizer.

Lightning can be damaging for two reasons, the danger of electrocution, and damage to property by physical disruption and fire. Electricity from lightning strikes will flow along reticulated water piping (not from water tanks, unless it be a direct hit) or the wires of telephone land lines (not mobile phones) or even fence lines. I heard of a farm gate being welded shut by the passage of an electric charge down the fence caused by lightning. It is unwise to take a shower or use a telephone land line during a thunderstorm, and be sure to turn off and pull the plugs from all electronic equipment (computers, TV), and in severe cases anything with an electric motor. If confronted by a fallen power line or a tree that has been struck do not run or even walk away, just shuffle, little bit by little bit, with the feet always fairly close together. This is because electrocution can only occur if there is an electrical potential difference. The amount of charge that is in the ground diminishes further from the source, and feet that are spread widely apart will each strike bits of ground with different strengths of electric charge, and the difference will go up one leg and down the other and electrocute you. Reassuringly, we know that an electrical charge resides on the outside of a hollow conductor, so inside our vehicles or houses we are relatively safe from electrocution, as long as we don't put a foot outside the door.

The second impact of lightning strikes is that they start fires. The air column directly surrounding a lightning flash reaches a temperature of around 30,000 degrees Celsius, or more than five times as hot as the surface of the sun (the corona of the sun is much hotter than its surface). Steel melts at 1,370 degrees Celsius. This tremendous heat causes any water or sap in the struck object to instantly burst into vapour, hence the explosive damage to those objects. Trees that are struck may violently shed their bark as the sap under it vaporizes and burst into flames, and similarly open grasslands, houses, boats, etc. may be set alight.

During electrical storms power supply interruptions may occur for long periods, in which case an installation such as a transformer has been directly struck, or power lines have fallen due to trees blowing over them, or be quite fleeting, due to a strike close by a power line that momentarily draws electric current away from the line by induction.

Volcanoes, earthquakes and tsunamis

These phenomena are of the deep inner earth and the ever-changing condition of the land surface, and although on-going, are not in any way implicated in the current process of global warming. However, it is important to understand them in order to be able to exclude them from involvement. Volcanoes influence the atmosphere over long periods of time by producing water vapour, carbon dioxide and sulphurous gasses, as they have done since time immemorial, but they are not responsible for current climate change. Volcanic events have the potential to greatly disrupt man's activities and even pose a threat to human survival, there are many precedents. This section describes some of the impacts these phenomena have had so far and possible future impacts. They are amongst Mother Nature's deadliest events.

Most of these occurrences are caused by movement of the earth's crustal plates. Many millions of years ago there were two super-continents, Gondwana in the southern hemisphere and Pangaea in the north. The revolution of the earth produces slow currents in its molten outer core, and these have caused the two super-continents to very gradually break up and bits of them to move about, creating "tectonic plates". Tectonic plates are slabs of the earth's crust that move very slowly over the planet's surface. The movement of these plates is driven by mid-oceanic volcanic ridges that run mostly north-south along the length of the major oceans, on the sea floor. They produce lines of up-welling lava which apply continuous pressure outwards to each side, a process known as sea floor spreading. This process also occurs on land, but this is uncommon. Interestingly, the Earth is the only rocky planet known (so far) to have plate tectonics.

Over a more recent span of roughly 50-90 million years the moving crustal plates of each hemisphere began to interact with each other, some of them colliding, others sliding past each other, and some moving apart. The plates are stationary for long periods because they snag together and cannot move against each other, but the pressure behind them is continual and remorseless. Over time, tension builds up at their interface and eventually causes them to move suddenly.

In the collision zones, one plate is forced under or over the other causing earthquakes to occur, often with accompanying volcanic eruptions and tsunamis. Where a plate is being forced down into the crust the sea is very deep, and is known as a "sub-duction zone" or a Trench. The Mariana Trench east of Guam is the deepest place on the planet, 11,000 metres deep, with a water pressure 1,000 times atmospheric. It is so large and deep it could engulf all of Mount Everest, and its peak would still be over 2,000 metres below the surface. Plate collisions have pushed up huge mountain ranges such as the Himalayas, which now have the only mountains in the world over eight kilometres high (14 of them). Mount Everest is 8.85 km high. The Himalayas are a special case because the two colliding continents, India and Eurasia, had similar rock densities and neither plate could slide under the other, so the thrust was upwards. The Himalayas are still rising at about a centimetre per year. Papua New Guinea is similar.

In the sliding zones, regular earthquakes are a feature, some of them severe, but there are no volcanoes or tsunamis. The San Andreas Fault is a well-known sliding fault and has wreaked earthquake havoc in San Francisco and Los Angeles. Plates that are moving apart occur in the Great Rift Valley, also the Danakil Depression of Ethiopia, which is a highly volcanic extension of the Rift Valley, and in Iceland. The island of Madagascar was separated from the continent of Africa by this process.

The Australian continent lies on the Australian plate, which is moving steadily to the north-east at about 7.5 cm per year (the old Indo-Australian plate is now recognized as split into the Indian and Australian plates). This is one of the most complex crustal plate collision zones in the world. Australia collided with the Eurasian plate (and two of its sub-plates) to its north, which is what threw up the islands of Indonesia, Malaysia and Myanmar. This collision zone has many highly active volcanoes. New Guinea is on the same plate as Australia, but it and other islands to its east were formed by collision with a complex of plates broadly linked to the Pacific plate, one of which is the Sunda plate. To the east, Australia is also colliding with the Pacific plate, and this has resulted in the formation of the Solomon Islands, Vanuatu, Fiji, and slightly earlier, New Caledonia. New Zealand is also involved here but has a more complex origin. These collisions cause stress to build up within the 25 km-thick upper crust of Australia, and the build-up of pressure can cause earthquakes within that land mass.

The "Ring of Fire" around the Pacific Ocean has several crustal plates in motion and in contact with each other, and there are many earthquakes and volcanoes constantly erupting along their margins. More than half the world's active volcanoes occur around this rim, for example in Japan, the Philippines, Indonesia, Papua New Guinea, the Solomon Islands and Vanuatu. Many volcanoes have produced pyroclastic flows, often referred to as the French "nuée ardente", which are instantaneous, explosive, massive eruptions of gas with an admixture of rocks and volcanic ash at a temperature of a thousand degrees centigrade and travelling at speeds of hundreds of kilometres per hour. They raze everything in their path, incinerating and instantly killing all animals including humans as they go, and can travel up to 30 kilometres in just a few minutes. Of all the disastrous events that occur in the world nuée ardentes are one of the most sudden and devastating.

There are about 20 very large "super" volcanoes around the world. Those that are fixed in one spot on the earth are called "hot-spots". They do not necessarily occur at crustal plate boundaries. Some, like Hawaii, are part of the mid-oceanic ridge system and some are terrestrial. Due to continental drift, the land mass of India passed over one of these on its slow 9,000-kilometre journey northwards. This hot-spot is in the Chagos-Lacodive Ridge, an archipelago in the middle of the Indian Ocean. As India passed over this hot-spot volcano it was ripped down the middle, creating the area of volcanic rocks in central India known as the Deccan Plateau.

The largest active hot-spot volcano in the world today is in the north of America, the foundation of Yellowstone National Park, which sits directly over a continental hot-spot. Only 600,000 years ago a huge eruption filled the area with lava flows, followed by a massive collapse that formed one of the largest land-filled calderas in the world (most calderas are water-filled) over 100 kilometres across which is Yellowstone, the world's first National Park. The volcano is quiescent today, and there are only the geysers and hot springs to remind us that there is a huge, simmering volcano underneath that beautiful scenery. At some future time, it will erupt again, probably catastrophically, killing many thousands of people and altering climate worldwide for years afterwards. It will certainly shake up the whole human race. It is, statistically, already overdue.

Some devastating volcanoes

The fabled and controversial ancient city of Atlantis, thought by some to be the world's first real civilization, is believed to have been destroyed by a huge volcanic eruption followed by a colossal tsunami. This occurred on the island of Thira, now called Santorini, in the Aegean Sea between Greece and

Turkey. This enormous volcano, one of the biggest in the world, produced prodigious amounts of pumice and volcanic ash, much more than has ever been recorded anywhere. The ash cloud had a global effect upon climate and caused worldwide famine for years. When the underground magma chamber was expended the volcano sank into it, and this now forms the largest marine caldera in the world.

In the year 1257 AD the biggest volcano known in recorded human history, Mount Samalas, also known as Rinjani Tua, erupted catastrophically. Samalas is part of a complex of volcanoes which includes Mount Rinjani on the island of Lombok, Indonesia. It was a truly giant eruption, and produced huge deposits of pumice, more than 8 times the amount that Vesuvius did when it destroyed Pompeii in 79 AD, and tremendous pyroclastic flows. Clouds of fine pumice dust went 46 kilometres high, four times higher than modern aircraft fly, and cooled the entire world. After the eruption and depletion of its lava dome, the volcano collapsed to form a huge caldera. A small volcano (an "anak" or child) has formed in the centre of the lake, called the Segara Anak caldera, with Mount Rinjani at its eastern edge. Volcanologists say that Samalas will erupt again, and as with other super volcanoes, it's only a matter of time.

Krakatoa was a big one. In 1883 it erupted explosively, and was one of the deadliest and most destructive events in recorded history. It destroyed 70% of the island and caused at least 36,500 deaths, partly due to a series of huge tsunamis. The sound of it was the loudest ever heard in the world, louder than the Hiroshima bomb blast, and anyone within 20 kilometres was instantly rendered deaf. It was heard in Perth and Alice Springs, and even up to 5,000 kilometres away. It is now undergoing rejuvenation as an "anak". Krakatoa is possibly the most dangerous volcano in the world. Vesuvius and Etna are also big ones, and the big ones tend to be less predictable. Etna is the biggest active volcano in Europe.

In more recent times, the volcano Tambora on the Indonesian island of Sumbawa was also a big one. It blew catastrophically in 1815 with a force equal to about 60,000 Hiroshima-sized atomic bombs and was the largest in recorded human history. The explosion was heard on Sumatra island, more than 2,000 kilometres away. It shot out red hot lumps of lava, some the size of houses, and an estimated 175 cubic kilometres of ejecta, including pyroclastic flows. The island's entire vegetation was destroyed, as uprooted trees, mixed with pumice ash, washed into the sea and formed rafts of up to five kilometres across. The eruption and the associated tsunamis killed an estimated 90,000 people, and many starved to death in the aftermath. The eruption was followed in 1816 by a "year without a summer" as clouds of ash encircled the world and cooled the whole earth by at least one degree. It is now thought to be extinct.

The super-volcano Toba produced one of the Earth's largest known eruptions, called a "mega-colossal" eruption. It occurred about 70,000 years ago at the site of the present-day Lake Toba in Sumatra, Indonesia. This catastrophic event was no ordinary eruption. It spewed an erupted mass 100 times greater than that of the largest volcanic eruption in recent history, the 1815 eruption of Mount Tambora in Indonesia, and was enough to create a six to ten years long volcanic winter, leading to massive die-offs of vegetation and the end of some species. It was followed by a prolonged cooling episode. This massive eruption changed the course of human history by severely reducing the human population. Toba and Tambora were probably even bigger than Krakatoa.

When Mount Pinatubo erupted in 1991 in Luzon it produced copious quantities of sulphur-rich gasses that reached 35 kilo-metres high, and by reflecting sunlight, cooled the earth by half a degree for two years. It also produced five cubic kilometres of volcanic ash. Heavy rain then formed lahar, a warm porridge-like slurry that washed off the volcano and slowly filled in the streams, completely buried whole towns, and flooded out across the paddy fields like a slow-motion tidal surge. You could run in front of it, but it would inexorably follow you. Despite a tremendous effort by the Philippine government by building huge dams and barriers they were powerless to stop it. How do you protect medium sized towns from complete lahar inundation which reached even over the top of their church spires, using bulldozers that don't work well in porridge anyway? With the heavy rain that fell after the eruption the whole area flooded, but volcanic ash had blocked surface drainage systems and urban roads became rivers. Later, when the lahar had cooled, there were people camped on top of the stuff because their land was down below, even if it was tens of metres beneath them. Nothing grows on lahar, it is mineral, just like angular coarse sand and quite porous, you could scour a pot with it but you can't grow carrots in it, it has not yet weathered into soil, and has no organic matter or microbes in it at all, and once it drains there is almost no water held in it to support plant growth; it's like an ultimate desert. Where the ash, or its mobile counterpart, lahar, was shallow (less than 30cm) there was an effort to scrape out holes in it down to the original paddy soil and plant tree crops, but that was sporadic due to little cultural knowledge. Many starved to death.

The big volcano at Rabaul on the tip of the Gazelle Peninsula in East New Britain, Papua New Guinea is one of the most active and dangerous volcanoes in Papua New Guinea. It had erupted in 1937, five years before the occupation by Japan, and entirely destroyed Rabaul. It erupted again in 1994 and once more devastated the city. Rabaul was situated on the flanks of the semi-dormant volcano that had collapsed after its last big eruption long ago to form a huge marine caldera. The caldera is surrounded by a steep volcanic ridge several hundred metres high which is the mountainous old volcanic rim. The airport runway was between two old volcanic vents almost at sea level with a smoking rubbish tip at one end, which unfortunately attracted birds – not ideal for an airport. In the recent eruption, Tavurvur and Vulcan volcanoes, which are on the edge of the former monster volcano, both erupted, destroying the airport and covering most of the town with heavy ashfall. There were only 19 hours of warning, but the city and most nearby villages were able to be evacuated before the eruption. Five people were killed—one of them by lightning from the eruptive column. Since then, the young cone Tavurvur located inside the caldera has been the site of near persistent volcanic activity.

The Taal volcano is a massive prehistoric volcano 60 kilometres from Manila, in Batangas Province, Philippines. It has erupted catastrophically several times in ancient prehistory. The expended lava dome eventually collapsed and formed a large caldera two kilometres across, known as Lake Taal. Since then a baby or 'anak' volcano has formed in the middle of the lake, called Volcano Island. This new volcanic vent is the second most active volcano in the Philippines, an island renowned for its active volcanoes. It shows frequent strong seismic activity and has erupted almost every year. One of its most devastating recent eruptions occurred in 1911, which caused great damage to agricultural crops and nearby buildings and killed many people. The detonation from the explosion was heard 970 kilometres away. Another eruption occurred in January 2020 with clouds of ash that swept over Manila, and massive volcanic lightning bolts.

Mount Mayon in the southern part of Luzon is a volcano that arises straight up from the coast, higher than Australia's Mount Kosciusko, a perfect, classical volcano like Fuji Yama or Kilimanjaro that constantly smokes but is unpredictable, and occasionally erupts lava and pyroclastic flows. It appears quite benign, but it is a big one. On one occasion when it did erupt, flowing lava, all the people ran for shelter to their church, crowded inside and were incinerated. Later, when the lava had cooled, all that was left of the church was the spire of its former roof. The whole church had filled with lava which solidified to basalt rock. The most recent substantial eruption was in early 2018.

I once saw an under the sea volcano amongst the Solomon Islands in the Pacific that emerged at times but was always visible as a red glow down below, with the sea actually boiling around it. Ships cannot float on boiling water, so they keep well away. It will probably form an island one distant day, that's the way Hawaii was formed. Of course, coral atolls and mainland outliers are entirely different kinds of island.

One of the present-day effects of volcanic eruptions is that ash clouds disrupt air travel, as happened in 2018 when Mount Agung erupted in Bali. Volcanic ash damages aircraft turbines. On one occasion, before this was understood, all four engines of a Boeing 747 were stopped by ash from Mount Galunggung in West Java, and within minutes an aircraft with four Rolls Royce engines became a glider. It fell 7,600 metres (25,000 feet) before the crew managed to re-start the engines and were able to land in Jakarta, averting a calamity. Afterwards, the passengers and crew formed the "Galunggung Gliding Club".

Earthquakes and tsunamis

Earthquakes are another hazard. The big earthquakes are related to movement of the world's crustal plates. The Richter scale is a base-10 logarithmic scale, it increases exponentially, so that an earthquake of, say 5 would be 10 times stronger than a 4, and 100 times stronger than a 3. The biggest magnitude earthquake ever recorded was in Chile in 1960 with a Richter reading of 9.5. Some earthquakes are very large, such as the magnitude 9.1 quake off the west coast of Sumatra which was 23,000 times more powerful than the Hiroshima atomic bomb and generated the 2004 Indian Ocean tsunami. It is thought that just to the south of Sumatra the old Indo-Australian crustal plate is breaking up and dividing that plate in two, which would account for such a large earthquake. This earthquake occurred along the Sumatra-Java subduction zone, where the Australian tectonic plate is moving underneath Indonesia's Sunda plate. This big earthquake moved the whole northern part of the island 20 metres to the north-east; you couldn't do that with nuclear power! This very large tsunami affected Sumatra and Thailand particularly, but several other countries as well. Over 230,000 people were killed.

Tsunamis are not displacement waves like ripples but shock waves, which propagate like sound waves by rapid successive compressions of the substrate, in this case water. The shock waves in water that produce tsunamis can move across the ocean at speeds of over 400 kilometres per hour. They are barely discernible in the open sea, but upon interaction with the sea floor they slow down, rise out of shallow water as a temporarily higher sea level and invade the land, and can devastate coastal areas.

Many earthquakes occur under the sea and are often accompanied by tsunamis. An earthquake of 7.5 magnitude occurred under the sea west of Sulawesi (formerly Celebes) in September 2018, followed by a devastating tsunami. I experienced a big earthquake in Manila, Philippines, which was 7 on the Richter scale at its epicentre, and was above 4 where I was on the 19th floor of a hotel. It threw me out of bed onto the floor. At first the movement was lateral, and the hotel shook from side to side, then it began to jolt vertically. All the high-rise buildings I could see from the window were swaying. I was told by a man who had experienced it that in an 8 it is impossible to run or even stand upon the ground, one cannot find any stable place to put one's feet. A little-known effect of severe earthquakes is that they can "liquefy" the ground, called a thixotropic effect. Strong tremors and vibrations can shake the soil into a sort of wet quicksand, and things upon it sink into the depths. A multi-story building in Luzon sank down vertically by one level, and later a new access door was put in at the new ground level on what used to be the second floor.

In some steep landscapes such as Papua New Guinea and Timor landslides are a natural event. Earthquakes often precipitate landslides and can shake large amounts of soil and loose rock down into valley floors, blocking drainage, often in several places at once. Water from the streams in these valleys builds up behind the piles of debris, and when more heavy rain occurs these unconsolidated "dams" break away and the pent-up water rushes downstream, engulfing villages and destroying infrastructure. Mudslides can be precipitated by quite small tremors or heavy rain, but many of them have pre-disposing factors such as clearing of the forest for agriculture.

The thing about volcanoes and earthquakes is not only the physical damage, it is also the noise, more frightening than the shriek of severe cyclones. It is a deep, growling rumble that shakes the ground and your very spirit, it comes from underneath, and it seems as though the whole world is coming to an end. Very large landslides are also frighteningly noisy, but the noise is more local in extent.

Running water and sea waves

R unning water is interesting in terms of its energy characteristics. It is controlled inexorably by gravity, as are satellites
and the moon. Flowing water possesses potential energy and
expends it as kinetic energy. Its energy is not like momentum
which dies away, it is constant, persistent and unceasing, like
gravity. The water in storage dams used for the production of
hydro-power has only potential energy because it is not moving,
but the actual execution of power production involves a conversion of this potential energy to kinetic energy as it is drawn down
the pipelines by gravity, and then produces electrical energy in
the turbines. Running water flows between places of potential
energy difference. A river flows from a multitude of small
detention areas in its catchment to the sea, the ultimate base
level, and whilst it is flowing between these points the water is
moving at close to perpetual motion, but it can only go forwards,
it can't go backwards against gravity. Nor can the moon.

But neither running water nor the moon can be seen as true
perpetual motion because they are driven by something, gravity.
In nature, the closest thing we have to perpetual motion is
"Brownian motion", observed in the early nineteenth century by
Robert Brown. He found that water consists of molecules, which
like all molecules are in constant motion. If there is a very fine
colloid in the water such as a suspension of small pollen or clay
grains these grains are in constant motion also, as they are continually bombarded by the molecules of water. The movement of
the water molecules is perpetual, it defies the power of gravity.
The grains themselves are suspended by this motion and can
never settle out. The water can be cleared by the addition of a
flocculating agent, alum (potassium aluminium sulphate) is the
best known. This causes the particles to aggregate into larger
clumps which soon settle out.

There is abundant energy in coastal sea waves, almost everlasting in that each wave is followed by another, continually. This is especially so when storms at sea throw up large swells. Wind physically pushes sea water into swells which move like ripples on a pond, but laterally, rather than radially. These swells propagate as displacement waves, each crest being pulled downwards by gravity to equalize with the adjacent trough, which is below it, it overshoots of course because it is a liquid and forms a new trough, and they all move on in sequence. Swells travel large distances towards the coast, gradually reducing in size.

The movement of such a swell/trough is not just surficial, it has a corresponding component downwards from the surface into the sea as well. When a swell approaches the coast the underwater part of it interacts increasingly with the upwards sloping sea floor, and the larger the swell, the further out to sea this begins. The "dragging" effect of this sea floor interaction causes the surface part of the swell to over-run the slowing lower part, and as this progresses the upper part rises and begins to fall forwards. The swell now has a more steeply sloping frontal face and the top "breaks" towards the shore, forming coastal waves. Surfers use the energy in these waves for sport and recreation. The energy in a wave is complex. It is no longer driven by the wind which provided the power originally vested in it, it now has residual energy of its own. This is a combination of kinetic energy due to its forward motion, and potential energy due to its height, until the wave expends itself upon the shore. The energy of this final impact can cause coastal erosion and retreat of the shoreline. However, if the under-sea component of the wave is substantial it can move sand on the deeper seabed towards the shore, providing material for later calmer wave action to complete the movement and replenish the sand upon the beach.

Where waves impinge tangentially upon a beach, they create a phenomenon known as "longshore drift" of sand. If a wave arrives obliquely upon the shore it moves beach sand upwards but in the direction the wave is taking, at an angle to the shoreline. Then the retreating water, having lost its forward momentum, drains directly down the beach towards the sea, taking this sand with it. The net effect is a gradual triangular movement of sand along the shore. The large sand islands along the central parts of Australia's east and west coasts were formed in this way.

On a more macro scale, the large volume of water that comes ashore in waves must find a way to return to the sea. Continuous successive waves prevent outflow of water through them back into the sea, and what tends to happen is that the beach becomes polarized, and areas form at regular intervals where this water does drain back to sea. These return-drainage areas are characterized by small or no waves because the water in them is moving away from the beach and has little impetus from seaward swells, and they appear as relatively calm water. They are known as "rips", and swimmers caught in them find themselves washed out to sea with this returning water. The survival trick is to swim sideways out of the rip, not back towards the shore against it, which can be very difficult if not impossible. When sea swells are large and continuous these rips cannot occur, and water drains back seaward underneath the waves.

We surf on waves only a couple of metres high although they can be ten times that. The maximum theoretical wave height that can be wind-generated at sea is calculated to be 198 feet (over 60 metres). Wow! Waves that big could sink large oil tankers or cargo or passenger ships. Large waves can have an almost vertical approach face and tend to break at the top. They blot out the sun. They must be terrifying. These waves literally shake the earth. They can be seen from outer space. Much smaller ones (25-30 metres high) have been detected seismically five thousand kilometres away. A big wave once shattered windows in the wheelhouse of the Queen Mary, over 27 metres above the sea (above the Plimsoll line), which were 25 mm thick Lexan which is akin to heavy reinforced plate glass, and flooded the wheelhouse.

The importance of scale

S cale is a much misunderstood but very important parameter, central to an envisioning of the mega-problems of the world and any understanding of climate change. Scale is a ratio, a tool for comparing things, usually in terms of magnitude, dimension or extent. It applies particularly to comparisons of things involving space or time, but it is not only to do with mathematics or the inanimate, it can be of anything; it can be used to compare aspects of political or emotional life; it is simply a comparison. In any comparison of data the items under scrutiny must actually be comparable, which means at the same scale. Scale is tangible, it can be readily observed and can be followed by anyone. The essence of its importance is, two or more parameters can only be considered simultaneously if all inputs are upon commensurate levels of scale.

Spatial quantities are easy to compare, and we can all deal with length and breadth and depth, we can all count, use a ruler or read a map. But in terms of scale, one cannot compare maps at scales of (for example) 1: 25,000 and 1: 250,000 because the levels of detail are quite different. A section of winding river can be seen as sloping to the north (say) on a map at 1: 25,000 scale, but on the smaller scale of 1: 250,000 that section may be just a small part of a general southward trend. It's the same with histograms and graphical representations. Anyone can find a section of a graph showing an upward trend or a downward one, but at a scale appropriate for the graph overall, neither of these might be a valid interpretation.

Scales of time are a little more difficult. With time, one thing is certain; time is omnipresent, it proceeds inexorably in line with quantum physics. However, in terms of scale, it is not necessarily an absolute, because we can review things in portions of time. Time can be considered at a variety of scales, small ones like last week or much longer ones like a hundred million years ago, and in this sense it is a variable. Each chosen unit is simply a subdivision of time. But in terms of comparisons, all inputs need to be at a similar scale. To compare events at different points in time may be fallacious; it is not valid to liken a week of freezing weather to the events of the last ice age, or to picture a dingo running with the dinosaurs.

When we look at scales of time outside the limits of human memory, we have a problem. The human mind is not a computer, perhaps it is in some ways far beyond computers, but scales of time take no account of that. For many humans the span of things is measured according to their own experience or living memory, and comparisons within that time frame are easy. However, measurements taken over much longer time frames are more difficult to comprehend. The variability in weather is clearly apparent in the relative severity of events such as storms, floods, droughts, landslides and fires, but the variability in global climate and other longer-term events such as volcanic eruptions is on a scale beyond that of a lifetime. The means of measuring these longer-term variables is well known to professionals but may be more difficult for those who see no perceptible change in them over the length of their own experience. Concepts such as light years of time, the extreme smallness of an atom, that matter is mostly empty space, the immense span of geological time, and the idea of a beginning or an end to the universe are difficult to contemplate on the scale of a person's life.

Scale is especially important in assessing climatic variables. Climate is inherently changeable, there have been immense fluctuations over geological time, and the particular span of it we are discussing needs to be made clear. The time frame that concerns us most is the time of the Anthropocene, which is within the last century or less. In any discussion of current climate change the data presented must be commensurate with this time period to be valid. Unfortunately, it is common to see climatic variables applicable over thousands or even millions of years used to denounce the validity of climate change as it is affecting mankind today. This practice is unscientific and quite misleading. The relevance of climate change to the human race is confined to the time frame of the Anthropocene.

Scales of space and time both apply to landscapes. In terms of time, the span of geological time is vast and comparisons within it need to take scale into account, entities must be compared at similar scales. A geological Formation consists of a sequence of Members, each of which can be very different. The mapping scale of a Formation may be about 1: 500,000 but that of a Member perhaps 1: 50,000. A particular Formation may be made up of several different sedimentary rock Members, another may be formed on volcanic rocks and would similarly be made up of Members but quite different ones. In terms of scale, we cannot relate the distribution of individual plant species, for example, to a Formation because each species has particular environmental requirements, we can only relate much broader taxa such as spinifex plains, savannah, desert, deciduous forest or tropical forest, which each contain whole suites of species that occur in association on a scale comparable to that of a Formation. It would be much more valid to map species at Member scale because a Member is more uniform geologically, but even then, the species would be unlikely to cover the Member uniformly. A Member would have variation within it due to local factors such as flatter areas with deeper soils, steeper areas with rocky outcrop, or different solar aspects, all on that one Member, and species distribution would vary with these, and so would the appropriate mapping scale of species.

Scale applies to human perception too. There are two common points of view dependent upon scale. One is that we are all different, which we are at an individual level in such things as eye colour, fingerprints and personalities, and the other is that we are all the same, which is also valid but over a much broader scale, that of the human population as a whole. Individually, we may have different body shapes and facial appearances, but as a species we all reproduce, feel pain, bleed, laugh, communicate, cry, and die. Genetically, all humans are 99% identical. And interestingly, we share at least 95% of our genes with chimpanzees!

The relative rate at which events occur can determine outcomes, and this can be an issue of scale. An intriguing issue of scales of time is whether the rate of atmospheric pollution by man is going to be faster or slower than the rate at which climate responds, specifically the time frame upon which each of these processes might come about. The outcome depends upon which factor is changing the fastest. If the rate of emissions is greater, then global warming and climate change are likely to intensify. Also, will the rate of global population increase be faster or slower than the escalating effects of global warming? Both are deleterious, but which will be the straw that breaks the camel's back? Another is whether coral reefs will grow faster or slower than sea level rise and survive or be drowned; and so on.

Sometimes figures cited for a quantity can be misleading. A comparison of the volume of a quantity may be given in one unit but the volume of the other quantity given in another unit, so that the one having the higher number appears to be the greater, which may not be true. For example, a particular volume may be cited in litres, such as 1,000,000,000 litres, another given in gigalitres (one), or Olympic sized swimming pools (400), or Sydharbs, the volume of water in Sydney harbour (0.002). In this case, the volumes are identical, but to those not familiar with scale the larger numbers appear to indicate a greater volume.

Issues of scale apply in politics also. Two common political obfuscations concern commensurate scales of time, and the relativity of numbers. Events projected to occur over a time of several years are often erroneously compared directly with those of immediate consequence. The numbers chosen in political arguments are also often at fault. Numbers that apply to a major economic sector such as banks or big business cannot rationally be compared with those that apply to a small business or an individual person's budget. A common one employed by politicians and the media is to cite one quantity in percentage points and compare it with a similar quantity cited in numbers, the numbers appearing to be much larger than the percentage, which is limited to a number of 100.

Precision is not the same as accuracy. Accuracy is an absolute term, and a measured number can be taken to however many decimal places one requires, for example a bacterium may be measured to many decimals of a millimetre, although in that case smaller units of measurement would usually be employed. Precision, however, is related to the item being measured, and the number cited must be in keeping with it. If one signposts the length of a steep grade on a road down to decimals of a kilometre, e.g. "Steep grade for 4.123 km" then that is getting down to metres, which is ridiculous in a slope several kilometres long, slopes on roads don't change that suddenly. If a dimension is quoted to the kilometre then that is the precision of it, a kilometre, and if given to the millimetre then it is precise to within one millimetre. The smallest unit of measurement in common use is the Angstrom Unit which is one ten billionth of a metre (10^{-10} m) and is used to measure electronic wavelengths and the size of atoms and molecules. A micron is one millionth of a metre or one thousandth of a millimetre, about the scale of micro-organisms like bacteria or the width of a human hair. Then of course we have kilometres, tons, and now Sydharbs, which are multiples of the volume of Sydney Harbour. The strength of a volcanic eruption or hurricane or other extremely powerful occurrence is often measured in "Hiroshimas", or multiples of the strength of that infamous bomb. The largest unit of measurement in common use is the Astronomical Unit, which is the distance from the sun to the earth and is used in astronomy.

The Global Impacts
of Man

The drivers of climate and weather

An understanding of how climate and weather work is basic to an understanding of how these atmospheric systems are now changing. Unfortunately for readers, meteorology is a very complex science and some of the following may seem to be abstruse. Climate and weather are not the same. Climate is consistent over long periods of time, but weather is to do with immediate short-term atmospheric conditions and is quite variable. Weather is what happens over a few or several days such as the time frame of Bureau of Meteorology forecasts, but climate is measured over time frames of from centuries to the span of recorded history, or even back into geological time. Where it recurs in a predictable way weather is considered to be a component of climate. Both climate and weather are driven by many factors, some global and some local.

Measuring change in a longer-term quantity like climate is really quite easy, but people (understandably) find it difficult to fathom. They say there have always been high and low points in climate so what's new? Wrong; the highs and lows they have in mind are in weather, not climate. Some peaks and troughs in weather have reached greater or lesser levels before today and some have set all-time records, but these are short term atmospheric events, not long-term climatic averages. Climate is assessed in a number of ways. Annual growth rings in trees, sequential coral formations, and stalagmites/stalactites are one type of indicator. Changes over time in polar ice deposits, the salinity of the seas, atmospheric concentrations of gasses such as carbon dioxide, and in the ratio of isotopes of elements like carbon are also used. These are averaged over many years; and of course, there is also statistical analysis. Climate change is assessed by the increasing frequency, escalation or decline, and persistence of all these events.

Globally, the weather is significantly dependent upon the circumpolar vortex or jet streams, which flow from west to east around each pole. This is especially apparent in the northern hemisphere due to the north-heavy arrangement of the continents. The only continental land masses that occur in the southern hemisphere, apart from Antarctica, are Africa below the bulge, Australia, and most of South America, all of which (except Antarctica) are slowly moving northwards. Jet streams are hardly ever uniformly circular in their orbit such as directly around a latitudinal circle, they have loops or meanders in them similar to the curves in a mathematical sine wave or a river in its coastal tract, where the wind flows alternately more towards the poles or away from them. These loops are not stationary, they precess or migrate around the poles so that they are alternately closer to or further from the poles, and they affect many countries sequentially. When a loop has formed towards subtropical regions cold polar winds are able to flow in behind it and occupy the area inside the loop, and countries in those areas have prolonged cold wet windy weather as a result. If the loop is towards the poles, warmer subtropical air flows to the inside of it causing a heatwave. These conditions continue for several days at a time until the loop has passed, and conditions are reversed. The speed of the wind in jet streams depends on the temperature differential to the north and south of them, they follow this temperature gradient, and wind speeds can double if the differential is large. The polar vortexes are a mechanism that transfers heat from equatorial regions to the poles.

In the northern hemisphere, southwards trending loops of the vortex which bring arctic air in behind them are most noticeable in Canada and North America. However, the same phenomenon occurs in the United Kingdom and Europe, curiously unrecognized, and often referred to as a return to winter, but it is the same circumpolar vortex. In contrast, the area along the equator (the inter-tropical convergence zone) has little wind, and in the old sailing days was known as "the doldrums".

Australia does not generate its own weather; it is influenced from all four marine directions. Each of the three oceans around Australia has a key climatic driver. There are no "oceans" to the north, the waters there are relatively small and are termed "seas". The main weather influence there comes from the annual north-west Asian monsoon. The Southern Oscillation (ENSO, or SOI) looks at the Pacific Ocean, the Indian Ocean Dipole (IOD) looks at the Indian Ocean, and the Southern Annular Mode (SAM) looks at the Southern Ocean, the former two being sub-equatorial. Australia's rainfall patterns largely originate in these oceans. The effects are expressed in the temperature of surface sea water and the atmosphere. The IOD and ENSO are characterized by alternating episodes of warmer and cooler surface sea water and fluctuations in atmospheric pressure, which tend to coincide, such that warmer seas are accompanied by lower atmospheric pressure, a situation that leads to increased rainfall. Global warming enhances these effects by raising air and water temperatures.

The IOD influences ENSO, episodes of which follow after a delay of about a year. This results (for example) in a higher rainfall condition occurring in the west of the Indian Ocean near Africa, but conversely so in the Pacific, so that wetter conditions pertain along the western coast of America. In 2019-2020 this pattern did occur, the effect coming from both oceans, and is what reinforced the severity of the drought and bushfire risk across Australia at that time. As well, the east-to-west trade winds and the Asian monsoon became weaker, exacerbating the drought conditions.

The Southern Annular Mode, also called the Antarctic Oscillation, is a measure of the strength or "storminess" of the westerly winds produced in the upper part of the polar vortex, and is an important driver of rainfall variability in southern Australia and New Zealand. A similar phenomenon occurs in the northern hemisphere, called the Northern Annular Mode, or NAM. It relates to the slow, seasonal, north–south migration of the trade wind belt that circles Antarctica and dominates the middle latitudes. This belt is characterized by a series of high- pressure systems that move from west to east across southern Australia. The highs alternate with low pressure areas between and immediately below them, but the whole system is an easterly movement of what is essentially a high-pressure atmospheric ridge. The south-easterly "trade winds" blow towards the north-west across the top of these highs. SAM has two modes. In positive mode, the westerly winds are gentler and produce calmer, warmer weather, the extremes of which are often termed "heatwaves". This is more common in summer and early autumn when the high-pressure ridge is further to the south. In negative mode it is the opposite, with cold, stormy, snowy weather dominated by a series of deep low-pressure systems and frequent cold fronts. This is more common in winter and early spring when the highs and lows are further to the north.

SAM can explain around 15 percent of the weekly rainfall variation in southern Australia during the winter and spring months which is comparable to what the Southern Oscillation Index (SOI) can, but SAM has a short attention span; episodes of it do not last for as long as the other climate drivers. The Indian Ocean Dipole can last for months and the SOI for years, but SAM generally only remains in one easterly-moving phase for a week or two.

Beneath the middle of the world's largest oceans are huge mid-oceanic ridges that have the highest mountains on earth; we cannot see them of course because they are under the seas. These ridges consist of long fissure volcanoes which span the ocean floors from south to north. Eruption of these volcanoes is almost continuous. The lava produced can raise sea temperatures and alter ocean currents, and so have an influence upon climate. These ocean currents affect climate rather than weather because the emanation of heat from these volcanoes persists over very long periods of time. The effect on ocean currents of slowly melting polar icecaps is also longer term; the difference is that the undersea volcanoes influence ocean currents by heat, but melting ice does it by influx of fresh water and worldwide re-distribution of mass.

Ocean currents strongly influence world climate. The Atlantic Meridional Overturning Circulation (AMOC) is a large system of ocean currents, like a conveyor belt, that carries warm water from the southern tropics northwards into the North Atlantic, driven mainly by differences in temperature. As warm water flows northwards it cools, and some evaporation occurs which increases the concentration of salt. Lower temperature and a higher salt content make the water denser, and this dense water sinks deep into the ocean. The cold, dense water slowly spreads back southwards, several kilometres below the surface. Eventually, it gets pulled back to the surface and warms again in a process called "upwelling" and the circulation is complete. This global process ensures that the world's oceans are continually mixed, and that heat and energy are distributed around the earth. This circulation is a major contributor to the climate we experience today.

Australia's weather is forecast from all these sources, plus of course a wide network of recording stations.

Global warming and climate change

It is important to recognize that global warming is not just a simple linear temperature effect; it is not just an increase in ambient temperatures by a degree or two impacting merely upon human comfort, which is a common but dangerous misconception; it is much worse than that. It involves significant disruption and intensification of all the complex physical and biological systems of the planet and the ways in which they interconnect. The power of nature is far greater than the power of atomic bombs, and if invoked can be much more devastating. Humans need food, fresh water and clean air, and with global warming all of these are at risk.

In the longer-term the progress of biological evolution has followed the changes in sea level caused by the coming and going of the ice ages, especially with the land bridges that formed concomitantly and allowed plant and animal populations, including humans, to migrate.

The earth's atmosphere is relatively thin. By comparison, if the earth was an apple, it would be (relatively) about as thick as the skin on the apple, an astonishingly thin and fragile film. Changes in this atmospheric skin influence global temperature. The earth is heated by absorption of shortwave radiation from the sun. Some of this is reflected from the earth's surface as longer-wave radiation in the infrared band. Nitrogen and oxygen, which account for about 99% of the volume of the atmosphere would allow this outgoing radiation to escape, but the "greenhouse" gasses, predominantly water vapour and carbon dioxide, intercept some of it and radiate it back towards the earth. Despite being present in small amounts overall, the natural blanketing effect of these gasses is what keeps earth's temperature stable and habitable for all its occupants, and small changes in their concentration can have seemingly disproportionate effects.

Water vapour is by far the most important of the greenhouse gasses and is responsible for most of the natural warming effect. However, the gas most affected by the activities of man is carbon dioxide, and most of that is due to the burning of fossil fuels. Not all of the world's carbon is in gaseous form. Rocks such as limestone, dolomite, marble, chalk and coral naturally lock away huge amounts of carbon as calcium carbonate ($CaCO3$), tens of thousands of times more than exists as atmospheric carbon dioxide. Natural, long term volcanic activity releases some of this stored carbon in zones where the world's crustal plates collide, and country rocks are drawn down into subduction zones and melted.

Worldwide, we are already seeing much of what was predicted years ago. The time frame of this is commensurate with the advent of the Anthropocene Epoch, or the meteoric rise in the human population of the world. What we see and record now is related to current weather conditions, and the time frame may be too short to indicate an indisputable change in climate. However, the changes are strongly symptomatic; they indicate a consistent difference in the frequency and severity of weather events and can be taken as indicators of a highly probable climate change. Climate specialists around the world are adamant that the climate is changing, and at a faster rate than has been predicted. Rates of atmospheric warming are now 100 times faster than they were before the Anthropocene. There is now more carbon dioxide in the atmosphere than there has been for the last 800,000 years. A recent Report (2019) in the peer-reviewed journal BioScience presents a warning based on data from many sources that the world is now facing "a clear and unequivocal climate emergency", but our political and corporate leaders continue to plunder the planet. A maximum increase in temperature of less than two degrees worldwide has been cited as an objective, but it must be remembered that some of the past ice ages and inter-glacials were precipitated by temperature changes of as little as four degrees.

The atmosphere contains many gasses, the most important being water vapour, nitrogen, oxygen and carbon dioxide (CO2), but also hydrogen, methane, nitrous oxide, ozone, chlorine and bromine (the halogens), and the inert gasses. The concentration of some of these can influence climate and global warming, plant life, fire, flood and drought. Global warming has been attributed chiefly to an increased concentration of carbon dioxide in the atmosphere, but methane and nitrous oxide also contribute. The burning of fossil fuels is the main contributor to the growth in atmospheric CO2, and there is an incontrovertible link between this and global warming.

Ozone is important because it shields life on earth from harmful ultraviolet radiation from the sun, but the ozone layer is thinning due to increasing concentrations of chlorine and bromine in the atmosphere. These chlorofluorocarbon gasses (CFCs) have been used as refrigerant chemicals and their release has caused atmospheric ozone "holes" near the poles, and as a result their use in refrigeration is now restricted. However, recent studies (University of Bristol; the Montreal Protocol) have found that recently increased emissions from this and other CFC sources are slowing the hoped-for shrinking of the ozone holes. In comparative terms, they were found to still have an effect equivalent to 25 billion metric tons of carbon dioxide, equal to all registered U.S. cars, which if averaged over 20 years equals the emissions of 270 million automobiles per year.

Recent studies (Bureau of Meteorology and CSIRO) show that there are definite ties between recent extreme weather events and long-term climate change. Globally, there has been a decade upon decade increase in temperature, with each of the last several decades being warmer than the previous one. There has been a poleward migration of severe tropical storms, which means further to the south in Australia, and they are larger, move more slowly on average, reach their peak more quickly and produce prodigious amounts of rain, all of which are thought to be related to an expansion of the tropics with warming of the seas. There is also evidence that these storms are becoming more intense overall, producing greater storm surges and bringing more rainfall and associated flooding on land.

Just as there is a water cycle, there is also a carbon cycle. The carbon cycle is complex, and changes in it due to the activities of man are difficult to assess. Contribution of carbon dioxide gas to the atmosphere is only part of the story, and with global warming the picture is changing rapidly. Carbon emissions come from several sources. Industry (including the manufacture of cement for concrete) is only one of them. Deforestation is another. When trees burn or rot they release hundreds of years' worth of sequestered carbon, and once they are dead they no longer capture and secure carbon dioxide from the atmosphere, so it is a two-pronged effect. Another function of tropical forests such as those in the Congo and the Amazon is that they release large amounts of water to the atmosphere as transpired water vapour which has a significant effect on world rainfall, and their demise affects climate. Drainage of wetlands especially peatlands is another contributor of carbon dioxide. Some significant sources behave in a state of flux, such as soils, which may emit large amounts of carbon, possibly more than industry does, depending on their exposure to sunlight, their temperature, and their organic matter content; or they may in some conditions sequester carbon.

Carbon sinks are the other part of the carbon cycle. The remaining forests are important sinks, as are the world's grasslands. The oceans have a natural, continuous exchange of carbon dioxide with the atmosphere and are the biggest carbon sink on earth, with marine plants such as seagrass and marine algae absorbing large amounts. Against this, increasing acid rain due to carbon dioxide dissolving in rainwater to form carbonic acid tends to acidity the oceans, which causes the release of carbon dioxide from marine stores to the atmosphere. Coral is a marine carbon sink, using carbon from the sea to build its calcium carbonate base. However, climate change is affecting coral reefs adversely, due to acidification of the seas and dissolution of its calcium carbonate foundation.

Methane (CH_4) is another carbon-based gas and is a potent atmospheric warming one, thought to be between 20 and 80 times as potent as carbon dioxide, although it is much shorter-lived. One source of methane is from areas of permanently frozen soil (permafrost) which thaw as global temperature rises. Permafrost covers about a quarter of northern hemisphere lands. It is frozen soil that contains ancient organic material from plants and animals, and as it thaws the organic materials decay and produce gasses including methane. In some of these areas the methane is so abundant that it can be ignited as it leaves the ground. The other gas produced as the organic materials decay, but over a longer time frame is carbon dioxide, especially if these soils are cultivated as they become warmer. Methane emitted into the atmosphere from oil and gas fields has recently been found to be considerable.

An important source of methane is animal agriculture, which is a huge contributor to climate change (United Nations). Ammonia is also produced, which contributes to the formation of acid rain. Domestic animals are estimated to produce about 20% of greenhouse gas emissions, which is more than all forms of transport including the world's cars, ships, trains and planes combined. Methane is the greater part of it. Methane is produced by the belching and flatulence of domestic animals, especially cattle. A reduction in ungulate (cloven-hoofed, cud-chewing) animals for meat consumption could significantly cut greenhouse gas emissions. A single cow produces more methane than a motor car, estimated at between 70 and 120 kg. of methane a year. A horse produces about 20 kg (kangaroos are close to that!), a sheep makes 8 kg, a pig 1.8 kg, and a human 0.12 kg, so a single human is really of little significance. However, the humans have the numbers; in 2019 there were 1.5 billion cattle in the world but 7.7 billion humans, a factor of 5, but even so, each human produces less than a kilogram of methane a year. Human flatulence contains not only methane but also hydrogen, and sometimes the smelly one, hydrogen sulphide (which is indicative of an unhealthy gut), all of which are flammable (H_2S is also highly toxic) and yes, it can be ignited!

Another important global warming gas is nitrous oxide (N_2O), which is about 10 times more potent than methane (and perhaps 300 times more than CO_2), although it is present in much smaller amounts. It is emitted as permafrost thaws, but the main source is from grazing animal manures, which apart from emitting large amounts of methane, also emit nitrous oxide. It is thought that with the green revolution and greater use of pasture fertilizers this atmospheric warming gas has increased by about 20%, over a time roughly equivalent to the Anthropocene, and is expected to double by 2050.

The climate outcome depends upon the balance of all these factors, and at present the over-riding net effect is that there is a significant release of carbon dioxide and methane gasses into the atmosphere, and it is driving global warming.

Water is the unifying element of all the world's natural systems. Water is the active component of the atmosphere; it covers three quarters of the earth's surface and is the basis of all life on the planet. Statistically, almost all the world's water (97%) is sea water. Of the small remainder (3%) that is fresh water, three quarters is locked up in polar ice, especially Antarctica, but also Greenland, glaciers, and groundwater. About 88% of incident rainfall is recycled to the atmosphere as evaporation, transpiration from plants, or soaks into ground-water reserves. Only 0.6% of the fresh water around us is captured and reticulated for human use. Gross supplies of reticulated fresh water around the world are diminishing, which is of obvious concern as water is the staff of life. In Australia, the main competing demands for fresh water are from agriculture, urban consumption, and ecosystem replenishment. Australia uses over 50 Sydharbs (Sydney Harbours) of fresh water per year, most of it for irrigation but also for mining, especially of coal.

Melting of the polar and other icecaps would cause sea levels to rise (or fall, if the icecaps expand) over time frames of from hundreds to hundreds of thousands or more years. Ice is very reflective, and a reduction in the area of ice at the poles would reduce reflectance of the sun's energy back into space and the earth would warm at an accelerating rate as a result. Complete melting of ice on terrestrial lands such as Greenland, the Arctic Ocean islands and especially Antarctica, all of which is fresh water, could ultimately raise sea levels world-wide by up to 200 feet, or more than 60 metres. This would dilute the saltiness of the sea in some areas and reduce its density and so alter major ocean currents, which would significantly influence both local weather and world climate. If floating sea ice, as in frozen seawater and marine icebergs were to melt it would have little effect upon sea levels, as floating ice already displaces an amount of water equal to its own volume if it were to melt.

The ice at both poles is currently contracting in area and getting thinner as well. The Arctic ice is quite shallow, only about a thousand metres deep, but the Antarctic ice is about three kilometres deep. The Arctic is currently warming almost twice as fast as the rest of the world, and since the 1970s about half the volume of sea ice in the Arctic has been lost. This is due not only to warmer air but also currents of warmer water melting the ice from underneath; there may be a pre-volcanic influence there. Already it is possible for ships to pass between the North Atlantic Ocean and the Arctic Ocean through the once-frozen seas and ice-covered islands of the Arctic, which cuts off a lot of sea miles and was not possible before. Antarctica is also one of the continents worst affected by global warming and is one of the fastest warming continents in the world. In 2020 the temperature there (Seymour Island) exceeded 20 degrees Centigrade for the first time, a record. New research shows that the Antarctic landmass is losing ice five times faster than it did in the 1990s, just two decades ago. Glaciers are also melting at unprecedented rates, some in the Andes have shrunk by 50% in just the last 30 years, and others in the French Alps have vanished altogether.

It seems to be popularly believed that a small increment of sea level rise won't do any real harm; what's a few centimetres upon the local beach, that's much less than normal wave height! Wrong. Ten centimetres of water across two thirds of the globe is a great deal of mass, given that one cubic metre weighs a ton. At 10 cm depth that would cover only ten square metres! What about the other two thirds of the globe that constitutes the oceans? How many square metres is that? And not 10 centimetres deep, but possibly up to 60 metres! It does not amount to a simple beach-side tally. A sea level change spans all the oceans, and with the associated global redistribution of mass could have a cataclysmic effect upon the land, much more than all our global nuclear bombs combined ever could. Bear in mind that the Pacific Ocean alone is bigger than all the world's land masses put together.

With sea level rise, there is potential for sea water incursion across marine barriers, flooding onto coastal plains, be they of fresh water or marine origin, increased height and reach of tidal bores, inundation of urban communities by backed-up fresh water or "new" sea water, serious coastal erosion and loss of beaches, flooding of surface drainage systems, damage to ports, railways, airstrips and roadways, many of which are low lying, failed access along our main communication corridors, depletion of marine life, more severe weather events, disruption of agricultural harvests, and destruction of infrastructure and crops by hail and wind. Even relatively short-term failures of fresh water supplies would decimate agriculture. Biological diversity, including in the oceans, will alter markedly. All this will be exacerbated during on-land floods due to heavy rain, coupled with king tides. Predictions are that we will see more frequent and increasingly severe events such as these. And who will profit? Crocodiles.

However, sea level incursion over the land may not be uniform around the world because land masses can change their level too, due to tectonic crustal plate movements. Some mountainous areas are rising at rates of 5-10 mm per year (Himalayas, Papua New Guinea, the Andes, Timor) which may currently be outpacing sea level rise, which is estimated at 3.4 mm per year. However, rate of sea level rise is predicted to increase to possibly 19 mm per year by the turn of the century. The concept of "relative" sea level rise provides a more useful measurement.

Sea level rise is already happening, and at an unexpectedly high rate. It is not only due to melting of the polar icecaps and on-land glaciers, because thermal expansion of sea water as it warms also contributes. One research group thinks that of the final possible sea level rise, Antarctica would contribute about 50 percent, Greenland about 25 percent, thermal expansion about 20 percent and glaciers about 5 percent.

Global sea level has already risen markedly, and in recent decades the rate has been accelerating. It is on record that sea level has risen by at least 30 centimetres in New York over the last 100 years, and there are similar records for other places. People in low-lying countries are having to re-locate because of sea level rise. In the Pacific region of Micronesia, the Marshall Islands, Kiribati and Palau are being slowly inundated by the sea, so are the islands of the Maldives, which are only two and a half metres above sea level and are predicted to largely disappear within 30 years. These changes are recent and are clearly measurable. On the larger land masses, as sea levels rise, water will not only visibly invade the land surface, it will penetrate behind the coasts via river valleys, around the back as it were, and occupy land areas from behind. It will also penetrate (and contaminate) shallow groundwater, raise phreatic water levels, and emerge from town drainage systems by welling up through drainage grates in streets and flowing onto roads, and this will be saltwater, undrinkable, and capable of causing corrosion and killing trees and other vegetation.

The oceans absorb much of the heat produced in the atmosphere by global warming. Now, the seas are becoming warmer, especially the shallower waters over continental shelfs, which are close to land and large populations of people. The revolving superstorms of the tropical climatic belt gain much of their energy from warmer air, and especially from warmer seas. A warmer atmosphere can also hold more moisture, which generates heavier and more prolonged rainfall, resulting in catastrophic flooding with associated massive landslides and environmental damage, exacerbated by the increasing denudation and exposure of large areas of soil by agriculture. Greater storm surges also occur, now coupled with an already higher sea level.

Higher sea temperatures lead to coral bleaching, which occurs in Australia's Great Barrier Reef when sea temperature rises above 30 degrees, and this is well under way in the northern part of the reef. Coral consists of tiny amoeba-like animals, anchored to a carbonate reef of their own making. Bleaching involves a loss of the coral's symbiotic algae which live inside the polyps, because survival of the algae depends on the temperature of the ocean water. These algae are photosynthetic and are a key partner for coral; it gets most of its nutrition from them, and the algae contribute the colour. With prolonged sea temperatures above 30 degrees the algae lose their ability to photosynthesize and are ejected by the coral. After about two weeks without these algae up to half the coral can die from starvation, and in this weakened state it can even get infections. High temperatures can actually cook the coral. The recovery time of coral is 10-15 years, but the recurrence of bleaching events has become less than that, about 7 years. These symbiotic algae have natural genetic diversity, but it is doubtful that variants able to withstand higher temperatures would become selected faster than the rate at which sea temperatures rise.

About 75% of the world's coral species occur around the multitude of small islands in Raja Ampat Regency (a municipality) of the West Papua Province, between eastern Indonesia and Papua New Guinea, known for its beaches, rich marine life and coral reefs. However, no area can be protected for long from the effects of global warming.

Coral needs sunlight, and water turbidity and deposition of silt inhibit its growth near the exit points of rivers to the sea, and in those areas, coral is naturally absent. However, increased turbidity and siltation caused by man's activities in coastal catchments is increasingly affecting previously healthy areas.

Contributions of sediment and nutrients from land clearing for the production of crops like sugar and beef, and coal mining, are the main culprits. It takes a lot more energy to dislodge a particle of settled silt than it does to transport it, so once it's down, it's there for good. Industrial pollution is also a problem, and as the Great Barrier Reef, the largest coral reef in the world, is close to the largest deposits of coal in the world, it is almost a choice between survival of the coral and the mining and export of coal. Around the world, 50% of coral reefs have already been lost from various causes and it is thought that 90% of them will be dead due to bleaching by the year 2050, regardless of palliative treatment of pollution and starfish predation. Coral reefs seem doomed to extinction, along with the magical ecosystems that depend upon them and contribute to the beauty and wonder of the human experience. Millions of years of the evolution of coral is being wiped out by the activities of man.

Some corals in other parts of the world are more robust and are adapted to sea temperatures warmer than in the Great Barrier Reef, and scientists are trying to create a "super coral" that might save the physical structure of the reef, although not all of its complexity and diversity. In the Persian Gulf corals survive at temperatures up to 36°C and it might be possible to install some of them on the Barrier Reef. The Barrier Reef is the largest reef in the world, with an area bigger than Victoria and Tasmania combined. The problem is a sort of marine version of re-afforestation on land because there is a similar issue of scale, the scale of it is immense in terms of not only the area to be covered with any treatment but also the length of time involved in rehabilitation. In the longer term, the effect may be palliative anyway because success of this venture is related to sea level rise; the question is, will such a new coral grow its calcium carbonate foundation at a rate commensurate with the rate of sea level rise, or not, and be drowned?

An ancient coral reef has recently been discovered off the north-west coast of Western Australia, running for about 2,000 kilometres between the Pilbara and Kimberley coasts (Global and Planetary Change journal). Cunningham and Bedwell Islands are on it. This reef is (or was) a western equivalent of the Great Barrier Reef and was at least as large; it is faintly visible on Google Earth imagery. It is estimated to have lived about 10 million years ago; the Great Barrier Reef is only about one million years old. Apparently, this ancient reef was slowly drowned as Australia drifted north and the sea floor subsided at a rate faster than the coral could grow upwards, and it was eventually deprived of sunlight and died. With global warming and sea level rise the Great Barrier Reef now seems to be in a similar position, but the rate of change is much faster. The Ningaloo reef is different as it is an "inshore" or coast-fringing intertidal reef and has largely escaped the warm ocean currents that lead to coral bleaching; but it will not escape sea level rise.

Increasing sea temperatures associated with global warming may affect not only coral. In Australia, due to strengthening of the East Australian Current, crocodiles are steadily moving southwards. Many fish and animal species are also re-locating. The highly venomous and sometimes fatal Irukandji jellyfish will within a decade be able to migrate southwards towards the most popular beaches in south-east Queensland; they are already present as far south as Hervey Bay. The tropical spiny stonefish has been found on the Sunshine Coast at Noosa. Increased temperatures and humidity are also favourable for voracious tropical termites (Mastotermes darwiniensis), and as they spread further south there is an increase in attacks on houses. The giant kelp forests along eastern Tasmania have already suffered a 95% decline due to warmer, less nutritious ocean currents.

Invasive and pioneer species may actually do better with climate change. There will be greater plagues of insects such as locusts, and of animals such as mice. More variable weather and increasingly crowded conditions will become more conducive to the spread of human illnesses such as influenza, Legionnaires disease, meningococcal infection, Ross River fever and other mosquito borne diseases. A global pandemic could be catastrophic. Coastal property values are likely to plummet by large amounts of money.

With the now well acknowledged process of global warming in place we are entering an age of "megafires". Fire has been related to climate in that it mostly occurs in drier or more seasonally dry climates, but this is now changing. Bushfire specialists are now proposing a new era of fire, which they term the Pyrocene. Megafires can only exist because plants exist, without plants there would be nothing to burn. Man-made things also burn, but except for isolated house fires, they require fire from burning plants to set them alight. With global warming, coupled with an increased atmospheric content of carbon dioxide, which is a major plant food, there is expected to be an increased production of vegetative material, so an increased quantity of flammable litter would become available to be burned. Fires will be worst where the fuel load is highest, and the lushest vegetation occurs in the more heavily forested higher rainfall areas. Protected areas in parks and reserves that are largely unmanaged are also at risk. In Australia, these areas are mostly on and east of the Great Dividing Range in Victoria, New South Wales and Queensland, where the bulk of the population resides. Humans are the only animals that can start fires, and in these areas they do. Severe electrical storms are common in these areas, and dry storms start fires. Episodes of hot, dry, windy weather and low humidity, as occur during severe droughts, are a prescription for catastrophic fires, exacerbated by shortages of water to extinguish them.

The worst of these fires are not just crown fires where the canopy of the trees catches fire, but fires that burn everything in their path, decimating masses of wildlife at every level in the ecosystem. Even rainforests are not immune; megafire weather can dry the soil surface under these forests and make the topsoil water repellent, inhibiting penetration of rainwater and drying out the trees till even they become susceptible to these monster fires. For years, fires have raged across the Amazon, particularly in areas felled and burned for pasture in order to graze cattle; so the photosynthetic sequestration of carbon dioxide is being changed to the release of carbon as the trees burn or rot, and the production of methane gas by cattle.

These are fast and furious fires. Really bad megafires such as those that have already occurred in California and Australia have been horrendous. Flames soar upwards for tens of metres, way above the tree canopies, and they can "spot" ahead for 5-20 kilometres. On the fire front they are an inferno, a seething, ferocious, maelstrom of fire. They are so hot that they melt metal, and they not only burn all the trees, green or not, and kill them and everything else, they have even burnt the organic-rich top layer of the soil. Some have swept right through small towns and villages including urban areas incinerating everything in their path, buildings, cars, people, bitumen roads, the lot, such as the ones that occurred near Athens in Greece in July 2018 and in the town of Paradise in California in November 2018. Another often serious effect of major fires is that a combination of raised dust and bushfire smoke can invade populated areas and cause respiratory distress in susceptible individuals. If heavy rain follows fire, sediment, ash and solutes can wash into reservoirs and pollute drinking water reserves. An unwelcome effect, as was experienced recently (2018) in southern California, is that megafires that totally destroy the vegetation and are followed by torrential rain can precipitate massive mudslides.

These intense fires create their own local weather system. Hot air rising from them creates strong updrafts, and currents of surrounding air are drawn into the base of the fire to replace it, increasing its intensity. The rising air contains not only smoke but moisture from the burnt material, and this can cause storm clouds to form and produce dry lightning, igniting new fires. Violent winds and fire-tornadoes intensify the fire and produce a blazing "fire-storm" which swarms across the whole landscape. Such a fire leaves a veritable moonscape, and large areas of it, and causes tremendous loss of wildlife. Anyone caught in the open by such a fire will assuredly die. Opportunistic back-burning and massive aerial attack may help control these fires, but they are almost impossible to extinguish. This is not like war between humans, this is the raw power of nature. Megafires are now more frequent and cover larger areas than ever before.

Radiant heat can kill ahead of the fire front. It is not actually heat, that is a misnomer, it is invisible electromagnetic energy that has no heat of its own, but when it strikes an object, unless it is reflected, the energy is converted to heat and can then move by conduction and convection, which can ignite a house ahead of approaching flames. If caught at home the best protection is by being inside the house because this frequency of radiation cannot penetrate solid objects, but if a person is caught outside, a cover of scant clothing is not enough.

There are many examples of more severe weather conditions. Whilst each could be regarded as just another episode of bad weather, collectively, they form an undeniable pattern. The recent decades have been the hottest on record (World Meteorological Organization, 2019). The year 2016 was the 40th consecutive year with an above-average global temperature, and extraordinary heat was experienced right around the world.

On the 13th of September 2017 England had its hottest September day in more than a century (34 degrees); climate change was cited as the main factor driving the record-breaking heat. There's almost no snow left on top of Mount Kilimanjaro, and there's even been snow in the Sahara Desert. Deserts, which already occupy one third of the world's land area, are expanding. In February 2018 Moscow had its biggest snowfall ever, a full winter months' worth in just 36 hours.

According to the State of The Climate Report (2018) by the Bureau of Meteorology and the CSIRO, Australia's climate and the oceans around Australia have warmed by just over 1°C since 1910, leading to an increase in the frequency of extreme weather events. There has also been more time in drought, and an increase in intense heavy rainfall across parts of the country. In Australia, the Spring of 2017 saw unprecedented heat waves across Australia and record high temperatures, some exceeding 45 degrees. The summer of 2019-2020 was worse, with figures of 48 degrees being common, and temperature maximums, averaged across the whole country, rising above 41 degrees. Australia's "summer" season is now a month longer than it was in the 1950s, and winters are correspondingly shorter. This sort of weather has been accompanied by more severe east coast lows and increased coastal erosion; more frequent damaging storms with high winds, intense rainfall and flooding; heat waves with periods of extreme heat; droughts with catastrophic fire conditions and intense bushfires; and altered patterns of rainfall distribution.

Flooding around the world has become disastrous. England had its wettest autumn ever in 2019 with widespread destructive flooding, and then in February 2020 storm Dennis brought a month's worth of rain in 36 hours and caused serious wide-spread flooding. Japan had the worst floods and landslides in centuries in 2018, followed by an even worse event in October 2019 when typhoon Hagibis catastrophically flooded large areas of the country, causing widespread disruption and significant damage. In December 2019 part of New Zealand had three times its monthly rainfall in as many days, and at the same time, across the Tasman, Brisbane had more rain in an hour than in the previous six months. South Asia has had its wettest monsoon season ever; one place in India had half the year's average rainfall in just five hours (2017) which caused serious flooding and loss of agricultural crops, and in 2019 much of that country was again under water.

Along the length of Australia, during January and February 2019, there have been catastrophic floods in North Queensland, and simultaneously, serious widespread bushfires in Tasmania, with severe drought across all the country in between. Over the Townsville district 1663 mm fell in one seven-day period. Ingham had 500 mm of rain in just 12 hours on February 2nd, 2019; other places had 150 mm in one hour, 200-300 mm in six hours, and half a metre overnight. One place had more than two metres in 12 days. The Ross River dam reached 247% of its maximum capacity (two and a half times!) and all floodgates had to be fully opened. Flooding was widespread, half of Townsville went under water. The drought-stricken inland also suffered, and thousands of cattle perished in widespread inland floods. This was estimated to be a one in 500-year event. And still it rained.

On February 12th, 2017 New South Wales had nine all-time essentially climatic records broken all in that one day. Sydney had its highest temperature in 80 years of 47.3 degrees Celsius on 7th January 2018; locally, temperatures went into the 50s. Broome has had a years' worth of rainfall in just one month, and Sydney had a full month's worth of rain in just two hours. In 2017 Western Australia, Queensland and the Northern Territory experienced their warmest winter since records began, and New South Wales and Queensland both had their hottest year on record, with multiple records being smashed. Across Australia, it is now common for a month's worth of long-term average monthly rainfall to occur in one single day, or even a few hours.

Severe drought conditions have hit Australia, also Zimbabwi, where even the native animals have been dying, the American West, and several other places. And it's not just seasonal drought. In Australia, it has been found that the amount of water flowing into the Murray River from all its catchments has almost halved over the last 20 years (Murray Darling Basin Review, 2020). Catastrophic, record fires erupted in eastern Australia in the spring and summer of 2019/2020, an early warning sign of worse to come. Sydney was enshrouded by smoke from bushfires 12 times maximum safe levels, worse than Beijing, and said to be unprecedented. These fires released significant amounts of carbon dioxide to the atmosphere, a positive feedback of global warming. Similar releases are occurring in the Amazon. The storm season of early 2020 was worse again.

The Caribbean Sea in the Gulf of Mexico has already seen an unprecedented sequence of superstorms, three of which (Harvey, Irma and Maria), all occurred in one season and were the most powerful Atlantic hurricanes on record, producing catastrophic damage and flooding accompanied by huge tidal surges and nests of tornadoes. Across the tornado belt of North America tornadoes are becoming stronger and more frequent. Hurricane Irma had the strongest sustained winds of any storm ever recorded and was the longest lasting superstorm in history (as at 2017). Hurricane Florence on the coast of Carolina (USA), super typhoon Mangkhut in the Philippines and Indochina, hurricane Michael, which hit Florida in October 2018, tropical cyclone Idai in Mozambique, and Amphan in east India and Bangladesh in 2020, are further examples. The enormous destruction that these super-storms can wreak has yet to be fully appreciated; wind speeds in excess of the upper forecast limit for category 5 hurricanes have already been recorded. Even outside the tropics, severe storms resulting from degenerating revolving superstorms have wreaked havoc, as occurred in Ireland in 2017.

One of the worst instances of mangrove forest dieback ever recorded globally struck Australia's Gulf of Carpentaria in the summer of 2015-16. The dieback, which coincided with the Great Barrier Reef's worst-ever bleaching event, affected 1,000 kilometres of coastline between the Roper River in the Northern Territory and Karumba in Queensland. The trees appear to have all died simultaneously. Losses were most severe in the NT, where around 5,500ha of mangroves suffered dieback. The Robinson and McArthur River coastlines lost up to 26 per cent of their mangroves. These mangrove deaths happened at the end of an unusually long period of severe drought conditions which prevailed for much of 2015 following four years of below-average rainfall, which was correlated with the strong 2015-16 El Niño event. This caused severe moisture stress in mangroves growing alongside salt marshes and saltpans. According to the journal of Marine and Freshwater Research, a combination of extreme temperatures with hot air and water, drought, and temporarily lowered sea levels were the probable cause of this dieback. The sea level dropped (locally) by up to 20 centimetres at the time of the dieback, when the mangroves were both heat and moisture stressed. Sea levels commonly drop in the western Pacific (and rise in the eastern Pacific) during strong El Niño years, and the 2015-16 El Niño was the third strongest recorded.

A global temperature rise is characterised in the short term by higher than usual temperatures in some areas and lower than usual in others, depending mainly on the location of the loops of the circumpolar vortex. As an indicator of climate change and a portent of things to come there was recently (2017) an intensely cold downwards loop of the arctic polar vortex over Canada and North America. Within it, those countries had the most severely cold winter's week for a century, with temperatures plummeting to around minus 40 degrees Celsius or colder in Ontario and Quebec. Some places had temperatures colder than at both poles, colder even than parts of Mars. In one storm, a "thunder-snow" event dumped 1.4 metres of snow on Buffalo City, New York State, in just 36 hours. In January 2018 a massive winter storm termed a "bomb cyclone" developed in central America in a southwards loop of the arctic polar vortex. Weather systems with low pressure are known meteorologically as "cyclones", high pressure ones are anti-cyclones. The term "bomb" relates to explosive deepening, or a very rapid drop in atmospheric pressure. These severe storms are intensely cold and have very high winds, and some of them revolve around an eye. These super-storms only develop within a loop of the polar vortex, which distinguishes them from typhoons and hurricanes which form over warmer, tropical seas.

Global warming and sea level rise appear to have been widely misunderstood. Most people have heard about climate change, however, what all the protesters and most of the public fail to see is that climate change is a symptom, a result, not a cause. If you ask them what the cause of it is, they will tell you that it is the burning of fossil fuel and release of CO_2, which is fair enough, although it's much more complicated than that. Then, if you ask them what can be done about it, and they will say, change to renewable energy sources. But replacing fossil fuels with renewables is only a palliative, a short-term effect; it should have been done long ago. The same goes for rampant, universal land and biosphere decline. If you ask, are the world's natural resources and environmental systems in distress? Fewer will respond because fewer have even heard of that because it is not as sensational, but those that do will say yes, and the cause of it is inadequate controls upon land use; some will implicate climate change. But no-one will point to the major, instrumental cause of both problems. The real, fundamental, underlying cause is the increasing effects of the burgeoning human population of the world, the huge and almost exponential growth of the global population, ravaging and polluting all the systems of the planet. That is truly the nub of it.

In human terms global warming is a slow, long-term process. It proceeds by gradual incremental changes, sometimes barely perceptible, which are easily accommodated in people's awareness and experience and go unremarked and are quickly taken as the norm - the boiled frog syndrome. People do not seem to understand the immense power of the atmosphere, the first revelation of which was that craft heavier than air could fly. This is not only in the potential of the huge westward-moving revolving super-storms but also in the eastward-moving thunderstorms, termed "super-cells", that track overland on a daily basis. These storms, with their heavy rain, strong winds, large hailstones and tornados have tremendous potential for damage and dislocation and are becoming more common and severe, and will get considerably worse as global warming escalates.

Claims that people, through their governments, can "adapt" to a warmer climate are fallacious, because climate change is not just a one-of event, it is an ongoing escalation. It has been postulated that the changes in climate already in place are likely to continue and worsen for many decades or even centuries before the world returns to its present state, even if the causes of global warming were to be removed immediately. Climate scientists around the world indicate that unless global emissions are significantly reduced now, this escalation of essentially meteorological events will continue and get worse. The initial impacts will be coastal, but not entirely. There will be a steady sea level rise and increasingly severe weather events, including flooding, droughts, dust-storms and fires, traffic chaos and closed airports. Despite all Man's technological advances, nature, in the form of global warming, will win in the end. If it is not nuclear war that annihilates mankind; dependent upon politics and time frame, it will be climate change.

Arguments claiming that the science behind global warming is divided into those who are for and those against are also fallacious. There are two kinds of 'science', the genuine, informed, rational and peer-reviewed kind, and the 'pseudo' or 'fake' science which is quite the opposite. Unfortunately, most members of the public, who have little scientific training, do not see the difference. Were this latter group to be disregarded, the scientific evidence that global warming is occurring and is caused by man is overwhelmingly definite. Global warming is not a matter of belief, it is a matter of scientific fact. The scientific findings on climate change are not just guesswork or estimates, they are actual measurements. Regrettably, at political level scientists are regarded as boffins with a barrow to push, rather than the voice of unambiguous truth. Worldwide, people pay more attention to celebrities, sports heroes, and the antics of political incompetents than to the world's most eminent meteorologists, internationally acclaimed experts and rational conservationists.

Politicians are not leaders, they respond to publicity, just as commerce does to money and religion to dogma. They have no concern at all for the welfare of the human race, it does not even cross their mind. Now, because of impecunious forward planning (coupled with a deficiency of intellect) governments let events control them, rather than them controlling events. That is largely because political figureheads are seen as celebrities, which is all that most of them are capable of being anyway, and in terms of qualifications requires only the usual assemblage of arms and legs. Try telling them that; they will respond with "Vote for me!". Some politicians still believe that the changes in climatic conditions being experienced are just part of a normal natural cycle and that there is no such thing as climate change, for example, the leader of the 'One Nation' Australian political party. Those who truly portend our condition are not the politicians, they are the scientists and allied people, and no amount of political rhetoric will make any difference to that.

Addressing climate change will involve a significant and rapid reduction in carbon-based emissions world-wide, which will require considerable change in global socio-politics and substantial upheaval and turmoil in every sector of humanity, as established systems change to accommodate it. In this respect political systems such as democracy (government of the people for the people by the people) may militate against change because people's capacity to act is limited by the state of knowledge and awareness of the voting populations they belong to. Despotic, totalitarian regimes will never do anything for the common good anyway. In 2018 the major polluting countries were China and the United States of America, and rationality in them might ultimately prevail. History has shown that much smaller changes such as the cost of living issue in France and Zimbabwe, demands for better wages in Bangladesh, a return to democracy in Hong Kong, escape from coronavirus lockdown in America and many other issues have created massive and often violent demonstrations, protests and riots, and the much greater adjustments required to accommodate climate change will be resisted vehemently, especially by the less-well informed, and will be a major impediment to mitigation initiatives. However, these ructions will be nothing compared with the primal catastrophe that unchecked climate change will eventually wreak upon life on earth, including all humanity. The worldwide impact of COVID-19 has offered a small insight, but a virus is a temporary thing and can be dealt with by man; climate change cannot. The world's intelligentsia are responding, but they have no hope against the tide of oblivious humanity.

A major impediment to change will be the widespread sybaritic, comfortable inertia of a Western world population that has grown up with the security, ease and expectation of life provided by an abundance of coal and oil. Many so-called 'civilized' populations now live in an artificial, sheltered world that is not in any way related to or integral with the natural systems of the earth. The ecosystem ethos has been all-but lost, and the human race has, unknowingly it seems, moved into a morass of artificiality and illusion, exacerbated by the technology-aided spread of misinformation and deceptively photo-shopped video clips on television and social media.

Human health is linked to population pressure. As natural animal habitats are pushed further and further back native animals are increasingly under pressure, and human interaction with them increases. This leads to human exposure to microbes such as viruses carried by native animals that might jump the natural ecological barrier to humans, perhaps mutate, and become transmissible from human to human, leading to pandemics. The other effect of loss of biodiversity is that it threatens our source of medicines, 50% of which come from the environment.

Whether people will have the restraint to not burn fossil fuels, when continuing to do so remains strongly advocated by politics and commerce, the means of doing so is familiar and extant, and supplies are stockpiled close at hand, remains to be seen. A transition from simply burning the obvious stuff (coal, oil and gas) to a whole new energy technology will put a huge demand on the flexibility and adaptability of mankind. Perhaps cynically, given human nature, that will never happen; humans will be unwilling, even inherently unable, to wean themselves off fossil fuels. Man may well be incapable of arresting climate change.

Australia has a misconception about climate change, it is that what Australia does will affect the climate in Australia; no, it is a global phenomenon and Australia's "in-house" contribution to it is minimal. Whatever we do (internally) will make little difference, except possibly to set an example and engender goodwill. Pacific islanders tend to blame Australia for their sea level rise problems, but they should be canvassing the major polluters, China and America. To illustrate this, Australia's Acting Prime Minister at the time (August 2019), said in relation to a Conference in Tuvalu that Pacific Islanders don't need to worry about climate change because thousands of them come to Australia every year to pick fruit! A senior Labor Party spokes-woman said upon questioning that Labor supports the mining and export of coal because it is an important part of Australia's economy! Politicians generally are uninformed about climate change (and many other scientific matters) and so continue to make such ridiculous and hypocritical statements.

Australia's global contribution to greenhouse gas emissions is low, about 1.3% of the total (2019), although it has close to the highest greenhouse gas emissions per person of any country. But the much bigger problem is Australia's export of coal and gas so that other countries can burn it, speciously exonerating us. A new report from research firm Climate Analytics (New York) notes that Australia is on a path to become the world's largest dealer of fossil fuels, and to become one of the world's worst contributors to climate change. Australia exports about 75% of its mined coal, which supplies almost one-third of the global export market (Australian Conservation Foundation), and in 2019 was the world's largest exporter of natural gas, a tripling since 2013. If all this fossil fuel were to be burnt locally, Australia would be one of the world's worst emitters of green-house gasses, along with China and America, but our politicians continue to support it in the name of export income and jobs. Of Australia's coal exports one third goes to Japan, the seventh largest emitter of greenhouse gasses, and one quarter to China, the world's biggest emitter. Up to 2019 China accounted for half the world's coal consumption and is planning large increases. And now Australia is contemplating drilling for new oil reserves in the Great Australian Bight.

The world as a whole is on track to produce more than 40% more oil and gas by 2040 than would be consistent with the Paris goal of limiting warming to less than two degrees Celsius (United Nations Report, 2019). The world is already beyond that goal.

At the United Nations Climate Change Conference in Madrid, Spain (December 2019), also known as COP25, hosted by Chile, Australia was ranked last in climate policy out of 57 countries and the EU, a rating of 0.0 and worsening, behind the second worst ranked United States with 2.8. It was also ranked among the bottom five of all countries in the world. This was mainly because of Australia's continued mining and export of coal, and the recent approval of the large Adani coal mine. Speaking at the Conference, teenager Greta Thunberg, Time Magazine's Person of the Year 2019, said countries are trying to give the impression that action is under way when "almost nothing is being done, apart from clever accounting and creative PR". She was publicly ridiculed by the President of the United States (Donald Trump). And still, politicians want to "strike a balance" and "get the balance right", when the issue is no longer political, it is an absolute; it is truly all or nothing.

Politicians try to turn every disaster to their own benefit; the economy, jobs and getting elected to government are all they seem to know or care about. Journalists fall in step, and add speculation, sensationalism, exaggeration, use of superlatives and selective reporting. Commerce, politicians and the media will seize any opportunity to sensationalize an event, the former for monetary gain and the latter two for power and personal acclaim. Their philosophy seems to be "never let a good crisis go to waste". Media news programs have become "shows" rather than objective reports. As an example, the visual media habitually confound the emissions issue by depicting images of large smokestacks emitting great billowing white clouds, which is not only misleading but ridiculous, because the problematic carbon-based gasses are invisible. Those clouds are steam, which is made of water; and the images mostly depict the dehydration of brown coal so that it can be burnt.

So far climate change is due to Man, and man can still reverse what he has done, or at least begin to limit the impact. Sir David Attenborough, a notoriously conservative and optimistic scientist, said (August 2019) that we are running out of time, but there is still hope. "Time" has been thought by the United Nations as being 12 years (2018). For man to even begin to address this man-made issue in a decade, even a generation, let alone getting to a level of net zero carbon emissions after a century of neglect, does not seem to be remotely possible. If the President of the People's Republic of China and the President of the United States of America, the world's two biggest polluters, think that climate change is ridiculous fiction, then what hope have we got?

Regardless of the changes wrought by man there is a looming disaster. We will soon reach thresholds, or tipping points, that will allow climate change to proceed on its own. As these are reached, the process of climate change will be removed from the influence of man and proceed under its own impetus, and will be self-sustaining. These points are fast approaching, and from there on, climate change will be completely beyond the power of man to retrieve. As just one example, consider permafrost. This is frozen soil, that occurs across large parts of the northern hemisphere. It is extensive, and contains large quantities of methane of organic origin, trapped inside the frozen soil. It will only take a final increment of temperature rise to cause the ice to thaw and change state to water, and then, suddenly, this tipping point will have been reached and the trapped methane will be released into the atmosphere. Methane is a global warming gas 20 times more potent than carbon dioxide. It won't all thaw at once, but every summer more and more of it will, in an accelerating positive-feedback loop. That release will be devastating; it will have been brought about by man, but once initiated it will be totally irretrievable, and there will be absolutely nothing that man can do about it.

Crossing one of these tipping points can increase the risk of others being crossed in a domino effect, they are inter-related, not independent, and one can follow another in sequence. The most obvious and immediate impact of population pressure is upon the atmosphere, and that in turn affects the oceans. There is a global linkage of ocean currents known as the Atlantic Meridional Overturning Circulation that mixes all ocean waters and distributes heat between the hemispheres, and strongly controls world climate. If the oceans warm, as they are doing, this circulation may slow down, reducing the transfer of heat from the southern hemisphere to the north. The unprecedented 2019-2020 drought and accompanying bushfires in Australia may be an early indicator of such a slowdown.

There are many such tipping points, and some of them are self-accelerating. Shrinking areas of sea ice reduce the reflec-tance of the sun's heat back to space and allow the oceans to warm, increasing the melting of the ice. Melting of areas of land ice such as the Greenland ice sheet would not only raise sea levels, it would significantly change key ocean currents and disrupt monsoon rains, critical for agriculture across much of the world. Loss of forests and their permanent conversion to savannah or grassland due to drought and fire is another one. Decaying and burning of forests around the world could cause these forests to begin emitting more greenhouse gasses (CO_2) than they absorb. When tipping points such as these are reached, and some may be very close, climate change will have grown legs, and will proceed inexorably on its own.

Perhaps paradoxically, it won't be the major impacts of climate change that will initially affect mankind, it will be peripheral and collateral effects that will impinge upon all the systems we rely upon in our daily lives and unthinkingly take for granted, and these could become significant with an even very small warming effect. Loss of things like power, home and workplace heating and cooling, telephone communication, credit card facilities, food and fuel supplies, piped water, transport of trucked essentials, movement of people, personal safety, medical help, freedom and justice, will occur, along with bushfires, destruction of property, fallen trees and power lines blocking roads, disruption of routine, general inconvenience, etc., at which stage people will start screaming to our social support systems for help. Then they will scream to the government to fix it, and if they can't fix it, they will go and vote for the Opposition party who will always claim that they can. And in the longer term, when things become catastrophic, they will scream for help to God almighty. But never, ever, will they stop producing more and more babies, or even recognize that overpopulation is the basic, underlying cause of it all.

Global warming is no laughing matter. It will inexorably escalate, and the effects will eventually be catastrophic. It is not applicable on a per country basis; no matter which country is contributing atmospheric pollutants the effect is global, what one country does, we all experience, it affects us all, indiscriminately. It could eventually put all of civilization, except perhaps for some enclaves, at mortal risk. Those enclaves may well backtrack towards the status of uncivilized or prehistoric man, as he was in the years BC. Global climate is changing. The climate monster is coming.

Classically, public concern is only generated after an unpleasant event has actually occurred, whereupon people ask, "Why weren't we told about this before? Why hasn't anything been done about it?" They want "everyone else" to tell them about it and do it for them. Ignorance is definitely part of the problem; people (especially politicians and journalists) tend to live very much in the present and think in narrow, insular terms. There are, of course, short term reasons why things happen, and they need to be addressed in that same time frame, because daily life goes on. But lamentably, people cannot see beyond them into the longer and much more brutal context of global warming, or even subdivisions of it such as rampant land resource and biosphere decline. The whole debate about climate change has revolved around the condition we currently find ourselves in, a changing climate and concomitant economic and societal distress, which is missing the most salient point; the abandoned, mindless overpopulation of the world by more and more humans, the single most important factor in the climate change debate, but one that no-one is prepared to discuss.

Denial of climate change is supported and even funded by those with vested interests in the status quo, especially the fossil fuel industry. In the public arena the climate change deniers are largely the less-well educated who claim that there have been droughts before and floods before and we will get them again as usual. By the time the nay-sayers understand global warming and become seriously alarmed it will almost certainly be too late. Why are we so reluctant to see it, and be warned? We experience ever more severe droughts, floods, fires and storms, and still man won't wake up. The changes are incremental over a long period of time, so the 'boiled frog' analogy seems to apply. The point being missed is that the climate change debate is currently about a symptom, climate change in isolation, not the real and admittedly complex causes of it. There is a chain of causes, including increased emissions of carbon-based gasses and severe and universal alteration of natural systems, but they can all eventually be sheeted home to gross over-population of the planet by human beings.

There is a caveat though, but on a less predictable, perhaps longer scale of time. There are a number of "super" volcanoes around the world, and if any one of them erupts catastrophically it could produce clouds of ash and sulphurous gas that would encircle the world and seriously cool global climate, perhaps even initiating another ice age. Two of the largest ones, Yellowstone in North America and Mount Samalas in Lombok seem likely to do so; there are several others. A direct hit by a large meteor is less likely but could occur, some have recently passed close to earth, one was closer to us than the moon.

World land resource deterioration

The world is beginning to recognise the perils of climate change, and we should now recognise another, hitherto unrecognised issue, that of the degradation of our land resources base, worldwide. The two are of course related. The continued deterioration of soils, fresh water and the biota contributes to that of the atmosphere and the seas, and should be of just as much concern. Since the industrial revolution, especially in the last generation or so, man has very significantly changed the whole nature of the earth's biophysical systems. Geologists have recognised this period as a new geological Epoch and have named it the Anthropocene. During this time the world population of humans has exploded. With that, there has been a huge increase in deforestation and land degradation worldwide, accompanied by pollution of two other global systems, the atmosphere and the oceans. The atmosphere is being addressed under the banner of global warming, and the pollution and over-harvesting of the oceans is receiving increasing attention. However, there has been very little recognition of the major problem of degradation of the world's land resources.

"Land" does not include everything. By definition it includes all of the terrestrial world, but not the atmosphere or the oceans. These are semantic omissions. The land/sea/atmosphere frontage relationship is dynamic and fragile. The seas interact intimately with the land masses and the atmosphere, and coastlines are the interface of all three. Contributions of runoff water from the land with their content of dissolved and suspended matter ultimately enter the seas, and atmospheric gasses such as carbon dioxide that are produced on land are absorbed or released by the oceans, and can influence the acidity of ocean water and affect its biodiversity.

All forms of resource deterioration are closely linked. Land degradation is a very significant contributor to global warming and climate change, and human welfare generally. Natural resources like soil and fresh water are as essential to man's survival as the air we breathe, and must be protected and used sustainably. The land and its soils are not properly appreciated, they are generally taken for granted, seen as an inexhaustible resource that has always been there and always will be, as enduring as the passage of night and day, just the dirt beneath our feet, and in no way in need of any form of management. Nothing could be further from the truth.

In its natural condition the components of "land" are closely related to one another. The degree of integration and inter-dependence of the flora, fauna, soils, and the broad open landscape itself is remarkable. The view that there are inter-relationships in the land between geology, soils and native vegetation that are predictable within a given climate was first outlined by Chris Christian, who was then Chief of the CSIRO Division of Land

Research. He contended that certain soils regularly formed on particular rock types, and the native vegetation that grew there was specific to those soils. It has been shown that these relationships are widespread and predictable in the same sense that the characteristics of biological organisms are related to those of their parents, hence the concept of "genetic" inter-relationships in land. This has become a fundamental precept in the systematic analysis of land.

In terms of land use, a basic principle is that all land should be used within the limits of its suitability. This includes that no land should be put to any form of use that diminishes its suitability for forms of use that it originally possessed, or precludes them entirely, such as good quality agricultural land being open-cut mined or committed to urban development.

Soil has lots of quite different geological substrates. Many of these are igneous, crystalline intrusive rocks like granite and gabbro, and extrusive ones like basalt and rhyolite. Sedimentary rocks like sandstone, shale, calcilutite and limestone, and metamorphic ones like marble, migmatite, quartzite and slate are less extensive. Raised marine deposits such as coral and chalk occur close to old coastlines. Over time, all of these rock types gradually decompose or "weather" to produce soil which is specifically related to each rock type, and which in turn will support a particular suite of native vegetation which thrives under those soil conditions, and which in tandem with climate creates a specific range of habitats for other biota.

On the land there are many processes at work, all of them interacting with others. Geomorphology, which is a study of the nature and origin of the land surface and the deeper regolith; ecology, which deals with the biosphere of plants, animals, insects, microbes and other organisms; and hydrology, which is to do with water and its movements upon and within the soil and deeper regolith (the weathered zone), are all aspects of these processes. Particular environmental conditions favour particular species, and once these conditions become known they can be used in reverse as biological indicators of prevailing soil conditions. As examples, in the Top End of the Northern Territory Eucalyptus papuana grows well on soils that have a plentiful year-round supply of subsoil moisture; stands of tall E. tetradonta do well on deep reddish sandy soils because they are a deep tap-rooted species; E. miniata thrives on shallow sandy soils as they have a shallow spreading root system; E. micro-crotheca grows on wetter heavy clay soils, and E. bleeseri on shallow gravelly ones. An area may be very dry seasonally but waterlogged during the wet season and support a monospecific stand of Melaleuca viridiflora. Other areas that are less distinct in their environmental characteristics might support mixed associations of maybe Eucalyptus confertiflora / E. tectifica / Corymbia clavigera, with E. oligantha on reddish structured clays. This sort of essentially ecological relationship is very consistent and predictable, in undisturbed environments.

However, an over-riding influence upon all of this is climate. Climate can be continuously wet, continuously dry or quite seasonal in its nature, and this exerts a profound influence upon all aspects of the land. Different climates will cause one geological type in different areas to produce vastly different types of soil, native vegetation, habitats, and all the dependent animal and invertebrate biota.

Significant changes in the landscape occur mostly during occasional catastrophic events such as severe cyclones, typhoons and hurricanes, climate change and climate boundary shifts, changing ocean currents, sea level changes, crustal plate movements and associated seismic events such as earthquakes, tsunamis and volcanic eruptions, and the rare impact of celestial bodies from space. Man ranks along with these events, and in a relatively short time he has radically altered the nature of the earth's surface. At other times the land mass remains essentially unaltered except for slow gradual erosion over geological time, both at the surface physically and internally by solution, and the gradual incision of drainage, all of which over very long periods can cause significant change. In the shorter term such things as flood, drought and fire events affect vegetative cover and this can whittle away at changes in the land. Often, as the land surface is slowly lowered by long-term erosion, remnants of the old original land surface may persist for long periods as plateau such as flat-topped hills, not to be confused with geological structural plateau, or gentle slopes above adjoining more dissected terrain. These reflect mechanisms of change in the land.

The pursuit of mining and the establishment of parks and conservation areas has little direct impact upon the condition of most of the world's land area; they have been likened to the area of a coin on the living room floor. The biggest impacts of man are in the huge areas of unregulated land, largely used for agriculture, grazing and human habitation. Inappropriate land management practices such as unwise clearing, cultivation, and overgrazing are widespread around the world and warrant much closer scrutiny. Everywhere, the natural environment is bulldozed to make way for the activities and structures of man.

Deforestation not only contributes huge amounts of carbon dioxide to the atmosphere, it also precludes further natural carbon sequestration. But deforestation also has a serious effect on the land itself. Tree canopies have a mitigating influence on the effect of raindrop impact upon the soil surface. The loss of the protective cover of the trees and the regular increments of litter that they shed leaves the soil exposed to the full impact of rainfall. This is exacerbated by fire, which consumes any surface cover that occurs. Tree roots also have a strong binding effect within the soil down to an appreciable depth, and their demise results in an increased susceptibility of the land to soil erosion and landslips. It is not uncommon to hear as I did in one country that further deforestation had been formally banned, but this was instigated only after all the forest had already gone and those in charge of forest conservation could no longer profit from its exploitation.

The clearing of tropical forests in countries such as Africa and South America is well documented. Belying the abundant luxuriant growth, and contrary to popular belief, these forests are not supported by rich soils. Many of the soils in these tropical areas are acidic and have very low levels of native fertility, the lush growth they support relies heavily upon nutrient cycling. This is brought about by the rapid decomposition of fallen plant and animal material by the very active biota, so that the minerals they contain can be taken up again by the trees and continuously re-used. When the trees are cleared these nutrients are immediately available for use by crops or pastures but they are not adequately replenished, and after just a few years the soils become infertile and impoverished. Reduced productivity results, but more importantly, soil erosion becomes a significant hazard. Within a decade, what was once a self-supporting tropical forest becomes a depauperate wasteland.

Forests provide about one fifth of the world's atmospheric oxygen as a "waste" product of the assimilation and processing of carbon dioxide during photosynthesis. But forests are not the biggest producers of life-giving oxygen. Silty sediments released from the world's land masses flow into near-shore shallow seas and allow huge populations of tiny phytoplankton to grow and multiply prolifically. This is a seasonal occurrence, and the areas involved are very large and encircle the world, they are so large that they can be seen from space. These tiny organisms carry chlorophyll and are photosynthetic, and collectively they produce much more of the world's oxygen than forests, amounting to more than half of all the oxygen in the atmosphere.

Marine algae such as kelp also photosynthesize and contribute oxygen to the atmosphere. With global warming, populations of these photosynthetic marine organisms are declining, and so is their production of atmospheric oxygen. Levels of oxygen in the oceans, too, are declining.

Soil erosion is an important component of land degradation. Many of the causative agents of soil erosion have been invoked by man. Removal of vegetation and its surface residues is important for two main reasons. It exposes the soil surface to rainfall, which causes detachment and mobilisation of soil particles, sometimes larger aggregates. Secondly, the eroded topsoil material takes with it much of the plant nutrient content of the soil including mineral nutrients and organic matter, reducing its fertility and weakening its physical structure, which lowers its resistance to more wind and water erosion and prejudices the future bulk of vegetative cover. The exposed subsoil materials are less conducive to plant growth and are generally more susceptible to erosion than topsoils and their exposure heightens the damage, with the formation of rills and then gullies.

Physical disturbance of soils also enhances the likelihood of erosion. Cultivation of marginal soils especially on steeper slopes is a prime cause of soil erosion. In hard dry soils it causes the production of large amounts of dust, which is tantamount to wind erosion. Overstocking steeper slopes can have a similar effect. Vehicular movement up or down slopes is also a prime contributor to erosion, especially in seasonally dry/wet climates. Traffic on pulverulent soils, which form "bulldust" when disturbed is another. Heavy vehicular traffic on soils that are moist can lead to compaction and reduced permeability.

Secondary salinization is now widespread. In drier climates such as Australia salt accumulation in lower-slope soils and the resultant curtailment of productivity has become a widespread problem. It is caused by the clearing of trees from the upper parts of slopes to encourage the growth of pastures there. Being deep rooted, trees take up moisture from well down in the soil profile and dispose of it back into the atmosphere through their transpiration stream. If the trees are removed this process is diminished, and because the pasture plants that replace them have a much shallower root system, infiltrating rainwater accumulates in the deeper subsoils. This water gradually moves laterally through the subsoil until it exits at the surface further down the slope, taking salt dissolved from within the soil with it. The water evaporates there, but the transported salt stays behind and accumulates, causing salinity. In more advanced communities this process is well understood. Drainage and leaching of affected areas does help but is expensive, and impractical where the problem is incipient and covers large areas. The longer-term solution is to replant trees on the upper slopes.

Acid sulphate soils have been found to be widespread across alluvial plains in many parts of the world. These contain sulphur-rich chemical materials called jarosite which were produced during the formation of these soils in sub-coastal environments, which were dominated by mangroves. Under natural conditions, these soils are continually waterlogged and exist in an anaerobic state, such that their acidity levels are maintained at or near neutral. However, large areas have now been drained for agricultural or other purposes. Drainage changes their anaerobic state to an oxygenated one, and the chemicals in them rapidly become oxidised and produce large amounts of sulphuric acid. This acid is soluble in water and leaches out of the soil profiles into the drainage network, where it renders the water unsuitable for irrigation, causes fish kills, releases toxic heavy metals, and causes erosion of concrete and steel structures such as bridges to the point of failure. Once the acidity inherent in these soils is released, conditions suitable for plant growth can only be reclaimed under experimental conditions; the land and its drainage water are rendered virtually sterile.

The off-site effects of land degradation are considerable. Deforestation, soil erosion, (etc.) not only have an in-situ effect, there are significant external impacts as well. Often, these imposed changes in catchments cause reduced detention and infiltration of rainwater into soils and an increase in runoff so that the time of concentration of runoff water in streams is reduced, and flash flooding in downstream locations results. On steeper slopes, retention of rainwater in the lower parts of the soil and in the deeper regolith leads to landslips and mudslides further downslope. Due to damaging land management practices in catchment areas, sediment and nutrient pollution of offshore coral reefs is becoming common. Mining activities, whilst they don't occupy large areas of land, can also have a significant off-site effect. Effluent from coal, uranium and other mines has already penetrated to groundwater, wetlands, conservation areas and offshore coral reefs. Fracking is another contributor of pollutants to both surface watercourses and groundwater. In many parts of the world the water in streams now carries toxic chemicals and dangerous organisms of disease, rendering it unfit for human consumption, or in some cases even for bodily contact. The physical condition of watercourses is affected too, not only by stream-bank erosion, but also the integrity of the water courses themselves so that natural patterns of water flow are disrupted, and with such ill-informed practices as the cutting off of meanders, increased velocities and greater peaks and troughs of flow occur downstream. Other off-site impacts are global and affect the seas and the atmosphere remotely, exacerbated by the effects of global warming.

Agriculture is of course an essential pursuit of man. Good quality agricultural land is scarce and all of it should be reserved for agriculture, but lamentably, much of the world's best agricultural land has now been irreversibly allocated to urban, mining and other non-agricultural interests. Farmers and graziers are a mixed lot. Analysis has found that only about 20% of farmers could be considered to be entrepreneurs or progressive, adopting advances and new developments in their field, and the rest simply do things as their father did, and badly, to their own detriment and that of the welfare of the land they farm; like father, like son. This 80% occupies by far the largest proportion of the land area, much of it marginal for agriculture, and the management of this part of our global land resource is largely in their incompetent hands. In this group, farming is not always carried out in accordance with the suitability of the land for each type of activity, and in many cases the land is subjected to impacts well beyond any concept of sustainability. Some large sheep and cattle stations do not fall into this category, but many do. Despite the rapidly increasing move in horticulture and aquaculture towards sheltered artificial environments and automation, degenerative broadacre farming and grazing remain a considerable part of the agricultural spectrum.

Both groups claim to be nature's conservationists. The concept doesn't really apply to the top 20% who are mostly on good agricultural land anyway, but in the case of the 80% it is quite clearly untrue, either in terms of impacts upon the natural environment or the simple maintenance of the soils under their custody. These farmers and graziers are the most malignant destroyers of land resources that the world has ever seen. They only conserve their own man-made environment, the one that suits their financial and emotional attachment to some piece of it and which helps them survive and maintain their way of life. They do not appreciate or protect the baseline environment at all. For them, the maintenance of income is regarded as of more importance than the conservation and sustainable use of the land. They persist with more and more land clearing, are responsible for habitat destruction, extinctions, plagues, weed invasion, secondary salinization, soil erosion, and disharmony amongst men; they exploit and ravage the land, they do not even attempt to conserve it, and because they have some sympathy or even adulation from the agricultural sector and are seen as the "salt of the earth" by an emotive public they consider themselves and their position to be sacrosanct. It has become a universal problem.

A further problem is that since time immemorial in many traditional communities, land areas have been passed on from one generation to the next as a cultural replicator. The age-old practice of dividing a patriarch's holding amongst his sons has led to successively smaller portions that require increased intensity of use, and abuse, to remain viable.

There is a widespread failure of governments to act or even to be aware of global land degradation; it is seen (if it is seen at all) as being the simple idiosyncrasies of the ground beneath our feet. Around the world there is not only ignorance about the land, but also corruption, self-interest at all levels and a lack of funds, and consequently there is little intervention or control. Like global warming, land deterioration occurs over a longer time term than that of immediate political reward and there are adherents and nay-sayers. The world is trying to overcome similar difficulties with other issues, so why not with global land degradation as well?

Ignorance about the land and application of inappropriate land management practices are almost universal, not only amongst farmers and graziers but passively acceded to in the mainstream of social and political culture. Around the world, land is increasingly becoming overgrazed, eroded, depleted of fertility, and is becoming progressively less productive. In the short term the changes are often subtle, quite evident to the trained eye, but the trend is clear. Soil material takes from thousands of years to many hundreds of thousands of years, perhaps millions, to form, and this time frame of replenishment makes lost soil essentially irreplaceable. Fertile agricultural soils are a very valuable resource, and it is imperative that they be used wisely and sustainably. Soil is the basis of civilisation, but it is very much a finite resource, and once it is lost, it is gone.

Conservation and pollution

There is nothing as truly pristine as the unfettered imprint of moving water, wind, or even volcanic ash upon the surface of the earth. Their patterns upon the ground are absolute, completely untouched by humankind; they are truly of nature in its most original and singular form. Particles of soil or other matter that are distributed by natural environmental systems leave a pattern upon the land. That pattern is, by definition, pristine. To see the swirls created by wind-blown fronds of spinifex in the desert with only the tiny tracks of an incumbent lizard overlying them, a clean new beach as the tide recedes, the growth of new grass, sprouts and flowers from the trees, tropical fish on coral reefs, the true calm of balmy sunshine and the fury of the storm are all pristine. It was said long ago, "Though every prospect pleases, and only man is vile" (Reginald Heber, 1826).

Our feelings about natural things, be they inanimate things such as the land with its soils and panoramas, or plants, animals and tiny creatures, are natural and innate within us, an inheritance from the intimate connection with the environment we once had so liberally and accepted as a matter of course. Humans have evolved in consonance with the real biophysical world and still need contact with it, for their emotional and physical health. Conservation and the wilderness concept are inextricably inter-connected.

However, burgeoning population, the advance of technology, coupled with the loss of volition imposed upon us by major commercial institutions and increasing regulation by government have significantly deprived us of this recourse. Some of us still live on the land or work in natural areas but many are now incarcerated in cities, with air-conditioned multi-story workplaces and dwellings, closed doors and windows, a balcony instead of a garden, and with perhaps only a pot plant to represent the beauty and grandeur of the habitat they are now remote from. Our instinctual yearning for this almost lost side of our humanity now seems to be displaced towards the "conservation" of natural systems in zoos, reserves, national parks and recreation areas.

The links with the wild within us have been expressed by aesthetic, creative and inspired people such as artists (the Heidelberg School of landscape painters); composers (Beethoven's Pastoral Symphony, Schubert's 'The Trout'); poets (Dorothea Mackellar's 'My Country', John Keats' 'To Autumn', Paterson's 'Clancy of The Overflow'); television presentations (David Attenborough); wilderness events (the Quilty); museums, ecotourism, and a plethora of narrative authors portraying natural beauty and environment.

Unfortunately, the biggest threat to the environment generally is man himself. Man's disconnect with nature is most evident in the cities. Just like other animals, insects, ants, bees, beavers, etc., humans strive to build their colonies (cities) and populate them to the exclusion of the environment around them. Increasingly, the natural environment is being seen and portrayed by urban dwellers as amusing, entertaining, diverting, quaint, rather than fundamental to life and truth. Loss of ecosystem integrity is as big a threat to humanity as climate change; we need clean air to breathe, clean water to drink, and uncontaminated soil in which to grow our food. We also need biodiversity. The United Nations warns that a million species of animals and plants are heading for extinction (IBPES Report, 2019). Because of the universal interconnectedness and interdependence of all species in the web of life there is no human future in a planet depleted of species. We must save the natural world if we are to save ourselves.

A very long-standing controversy has been the debate between commerce and the environment; the economics of natural resource management. This is a complex issue and fraught with vested interests (including political ones), ignorance, tendentious argument and sensationalism. Much of the failure to resolve environmental issues is due to the scant understanding most protagonists, including farmers, "greenies", politicians, journalists, and the general public have of science, especially environmental issues. The protagonists are usually ill-informed about the scientific facts or do not understand them, and the outcome is commonly awarded to the most vociferous. Science knows the facts and what ought to be done, but politics has the power, and frequently over-rides or sidesteps science except when it suits them not to; they change and select data to suit their re-election. Politically, the environment will always come second to social and economic factors, and only after those are satisfied will rational scientific evidence be considered. In the debate between science and politics, politics will always win, and the environment will come second, or third. Power without knowledge is ignorant, immoral and corrupt.

An instance of the agriculture/environment conflict relates to the Barmah Forest, along the Murray River. There is a spot along the river called the Barmah Choke, a natural narrowing of the tract that seasonally backs up river water and inundates the adjacent floodplain, which has, in consequence, a flood-main-tained forest of River Red Gums (Eucalyptus camaldulensis) now preserved in the Barmah National Park, which, together with the adjacent Millewa Forest constitutes the largest Red

Gum forest in the world (285 square kilometres). An argument developed over water allocation between farmers upstream of the Barmah choke who wanted water for irrigation, and advocates of "environmental" flows, who wanted to save the iconic forest from dying due to lack of seasonal floods. The Murray-Darling Basin Authority eventually released sufficient water to flood the Barmah Forest and save it from impending demise. Farmers upstream saw this water running past but were not allowed to use it, and were quite upset about it. Farmers downstream of the Barmah Choke were allowed liberal use of the water after it had passed beyond the Park, and those upstream were additionally indignant as a result. The debate was featured on television, hosted by a reporter from the Channel Nine Network's 60 Minutes program who was quite ignorant of the facts and sided liberally with the noisiest interest group, who claimed that the floodwater in the forest was a "deliberate waste" of water that they should have been allowed to use, and that it was killing the forest by drowning it – patently absurd, the forest was half dead due to the preceding years of inadequate flooding, not the life-saving inundation it was finally receiving. A following interview with the Federal Minister for Water Resources, Drought, etc. demonstrated that he was not only ill-informed but clearly incompetent.

This kind of partisan, vested interest dispute, fuelled by the media, with only mindless, inept government as adjudicator occurs in almost every case of competition for resources, and has done throughout history.

Man's activities strongly militate against conservation in almost every sphere. More and more, we see that man is in opposition to nature, rather than in harmony with it. Commerce and government are largely to blame; they are directly opposed to the natural environment and thrive on its exploitation, to the detriment of viable ecosystems. We are on the cusp of consuming more than nature can provide. Continual striving for expansion and development of the built environment to provide increased job opportunities and sustained economic growth are axiomatically in conflict with the natural biophysical world. Farmers claim to be natural conservationists, but they rarely are. The old 'shoot it, chop it down or burn it' syndrome is well and truly alive.

The entire trajectory of human development has been shaped by the availability of fresh water. The most important, key conservation issue on earth is going to be the availability of fresh water. According to the United Nations, globally, more people now die from unsafe water than from all forms of violence, including war. Waterways are often visibly contaminated by urban refuse and are completely unsafe to drink or even bathe in. All waters, including the seas, are universally polluted by the ubiquitous plastic bags, fishing nets, shipping containers and many other items. Fresh water supplies around the world are diminishing, and water in pipes is in short supply. Of the available fresh water in the world very little is reticulated or otherwise available to centres of population. Jakarta has been on severe water rationing (2018), Cape Town almost ran out of tap water, and other large cities have experienced water shortage and rationing. Water in storages for agricultural crops and farm animals is often in short supply.

Countries like Australia are often in drought, with the probability that it might become protracted, and in such a naturally dry continent fresh water may actually run out in some areas, including drinking water, the stuff that keeps us alive. If there were to eventuate a serious water shortage across Australia, how would the country cope? We can't truck it to the bigger centres, they need much more than that. Do we shift people to higher rainfall areas? Millions of people? Do we urgently dam wet areas and pipe water to the regions? How do we maintain agricultural production? This is going to be a serious matter, perhaps of life and death for some. Yet we continue to flush our toilets with drinking water.

Of Earth's total water (consensus from several sources):
- 97.2% is saline, in the oceans and inland seas
- 2.1% is in glaciers
- 0.6% is in groundwater and soil moisture
- less than 1% is in the atmosphere
- less than 1% is in lakes and rivers
- less than 1% is in plants and animals.

So, if the oceans hold about 97 percent of the earth's total water, only the remaining three percent is freshwater. Fresh water includes water in ice sheets like Greenland, the polar ice caps, glaciers, icebergs, snow, bogs, ponds, lakes, rivers, streams, groundwater, the atmosphere, and plants and animals. Of the world's 3% of liquid fresh water 99 percent is groundwater, much of which is not accessible to humans. Most of the water people and all other life of earth need every day comes from the remaining 1%, which is surface-water resources.

Large quantities of waste materials from the world's land areas have been washed into the oceans. The Great Pacific garbage patch, also described as the Pacific trash vortex, is an accumulation of floating debris in the North Pacific Ocean between Hawaii and California. It covers 1.6 million square kilometres, more than twice the size of Texas or three times the size of France. It has an exceptionally high concentration of plastic, chemical sludge, and other non-biodegradable debris that has floated from mainly terrestrial sources and been trapped by the large system of circulating ocean currents in the North Pacific. It is rapidly accumulating. There are five such accumulation zones, termed Gyres. The second largest is the North Atlantic garbage patch, an area of man-made debris floating within the

Sargasso Sea, estimated to be hundreds of kilometres across, and also bounded by several ocean currents. An estimated eight million tons of plastic pours into the sea every year. To contain it, plastic waste needs to be continuously recycled, never committed to landfill.

Beaches are also under threat. The United Nations Environment Program points out that residues created by humans are increasingly littering and polluting beaches worldwide. There is an increase in sargassum, the dead remains of algal growth (a large brown seaweed) that floats in island-like masses derived from the Sargasso Sea appearing on beaches globally; it is thought to be caused by pollution of sea water by agricultural and other chemicals, and climate change.

Many major cities around the world are now seriously polluted. It's not only water, but particularly in urban areas and extending into the country air pollution has become a critical human health hazard. In New Delhi, air pollution is said to be subjecting its citizens to the equivalent of smoking 40-50 cigarettes a day, and there are many runners-up, such as most big Asian and American cities including Jakarta, Beijing and Los Angeles, and most of those in Europe and Britain. Extensive use of electric motor vehicles is known to be one solution, and China in particular is strongly promoting electric cars as a means of mitigating air pollution in its cities, however, it must be remembered that a large proportion of the pollution due to cars comes from tyre and brake wear. In some respects China is unduly maligned, because as well as coal and oil it has enormous wind, solar and hydro power developments, it has to, to keep pace with a constantly increasing demand; however, little attention is paid to environmental concerns. Their dams on the Yangtse and upper Mekong Rivers are examples.

Man's impact on the biosphere has affected every animal, bird, fish, insect and plant, with habitat destruction, extinctions of species, inter-country transfer and proliferation of invasive species of all kinds, the spread of diseases, and soil erosion. Coral reefs are being destroyed world-wide not only by global warming but by blasting to kill and salvage fish, and mangroves have been extensively cleared to make way for commercial prawn farms. Sea waters in near-coastal areas, especially close to major cities have been likened to marine deserts, due to chemical and sediment pollution and over-fishing, especially of shellfish. Another issue is that native flying-insect numbers have been observed to have crashed to an alarming extent. Christmas beetles were said to be so numerous 100 years ago in the Sydney region that they could be found floating in the harbour in huge numbers at that time of year; now there are hardly any.

It is reported by the Kew Royal Botanical Gardens that more than 20% of all terrestrial plant species are now at risk of extinction. In places like the Amazon, the Congo, Malaysia and Indonesia tropical forests have already been extensively cleared to grow fodder for cattle and to make way for oil palm plantations. Provision of fodder for grazing beef cattle is the foremost cause of tropical deforestation. Ruminating cattle produce methane (mostly orally, from their rumen), a greenhouse gas many times more potent than carbon dioxide. Much of this methane emission could be saved by the consumption of chicken or fish, or non-ungulate animals such as Australia's kangaroo, even microbial or insect protein, instead of beef. In some areas, oil palm plantations have replaced around 80% of former tropical forest land, such as in Sumatra, where half the original rainforest has been lost to oil palm in the last 30 years. Commerce continues to utilise palm oil because it is cheap, and it is now found in very many of our foods. As the forests are lost, native animals dependent on them for habitat and food are also lost, such as the Asian elephant and rhino, orangutans, Indian and Malayan tigers, and many other species. The last male

Sumatran rhino has now died. The Australian Mountain Ash is the second tallest tree on earth, and a big one weighs more than the biggest animal that ever inhabited the earth, dinosaurs included, the blue whale; yet we happily fell these magnificent trees for woodchips.

A remarkable thing about all forests and even grasslands is that the combined weight of the microbes and arthropods that live on and in the soil is far greater than the weight of all the larger native animals living on and in it put together. Spiders alone, worldwide, are estimated to weigh 25 million tons (and to eat more than all humans, and could eat all the people in the world within a year!), and all the ants in the world are thought to weigh about as much as all the people on earth. However, it is estimated that domestic animals now outweigh the combined weight of all the native animals, and that domestic fowls are probably the most numerous birds in the world. How that compares with the massive bulk of humanity is anybody's guess, but if there are 7.7 billion people on earth and each weighs 60 kg, that totals 460 million tons, roughly 10 million whales or 80 million elephants.

Carbon dioxide in the atmosphere combines with water to produce a weak acid, carbonic acid, and this falls to earth in rain. Increments of this acid slightly raise the acidity of the seas, and may result in softening or weakening of the calcium carbonate exoskeletons of small animals like krill, foraminifera and other plankton and put their populations at risk. If the water becomes too acidic the eggs of krill may not hatch. These tiny organisms are vital food for filter feeding whales and the biggest of all sharks, the basking shark. Sponges, flamingos and manta rays are also filter feeders. Other marine animals and fish are reliant on plankton through the food chain. So regardless of some countries' insistence on whaling for "research" purposes, increasing carbonic acid in the sea may exterminate the whales anyway.

The saga of the cane toad is familiar to us all. It was introduced to save the northern sugar crop from beetles – sugar! A dietary poison! Cane toads can exude a highly toxic poison which kills anything that eats them; that's one way to deal with your enemies! Cane toads are causing devastation across the northern Australian goanna population, a very necessary top predator, and their possible 90% demise or perhaps extinction. In Arnhem Land the cane toad has killed many native animals and birds that have eaten them including almost all the water monitors, and unless controlled will eventually poison them to extinction. Now, both the toad and sugar are recognized for what they are, total undesirables. Then there is the recent importation into Australia of the red fire ant. This ant, if it proves to be un-eradicable or un-containable could become an environmental disaster that rivals many other imported species, including not only the cane toad but feral cats, foxes, camels, and feral hoofed animals including buffalo, cattle, deer, donkeys and goats.

Domestic cats gone wild kill more than one million birds, one million reptiles and one million small mammals every day. Across Australia, this means that feral cats are killing more than 2,000 native animals every minute (Australian Wildlife Conservancy, 2019). They are the single greatest threat to Australia's endangered wildlife, especially small mammals. One in three native mammals is at risk of being added to Australia's shocking extinction record, already rated as the worst in the world, with 31 mammals (10 per cent of our original mammal fauna) now listed as extinct. Between them, feral cats and imported foxes have all but decimated Australia's formerly widespread bilby population. The feral cat problem can only be made harder to address if it is added to by domestic cats being allowed out at night to plunder wildlife, and their continual defection to the wild. It is estimated that roaming pet cats kill around 150 native reptiles, birds and mammals per year, each, which means that across Australia they account for maybe 390 million additional deaths every year. There appears to be no reason why domestic cats should not be required to be registered with the local Council and sterilized, as is already the case for dogs.

An estimated 4,000 koalas are killed every year by dogs or run over crossing the road; more die in bushfires because they cannot escape. Birds such as the Indian Mynah have taken over the habitat of native birds and reduced their ability to find nesting sites for breeding and survival.

Worldwide, tilapia and carp are amongst the top 100 most invasive fish species. They are regarded in much of the world as satisfactory eating fish and are popular with some anglers, and tilapia are widely farmed in America. Tilapia flesh is rather bland, and carp frequently taste 'muddy' because of their bottom-feeding habits. In Australia they are regarded as pests in natural aquatic ecosystems because they reduce water quality by causing turbidity, and by competing with native fish for food and space. Carp were introduced first and are now the most abundant large freshwater fish in the Murray Darling Basin and the dominant species in many fish communities in south east Queensland, where they are a significant threat to native fish and their habitats.

Across northern Australia large areas have been decimated by feral buffalo. These animals were introduced in the early 19th century and abandoned to the wild in the late 1940s. Their natural habitat is wetlands, and they found northern Australia very much to their liking. They multiplied and spread rapidly across the Top End doing serious damage to the environment. The views you see in the media, designed to boost the tourist industry in such areas as Kakadu National Park are often carefully selected or photoshopped, and misleading. Buffalo not only destroy wetlands and the essential habitat of so much wildlife, they also damage Aboriginal rock paintings by rubbing against them, and they can even ring-bark large paperbark trees by rubbing and by chewing the bark. They also cut channels to the sea through coastal dunes and let saltwater in (some of the coastal plains are actually below high tide level, protected by the dunes) killing whole freshwater ecosystems and even large trees. Culling and domestication of tens of thousands of them has been undertaken in the brucellosis eradication campaign but the nucleus is still there and culling alone will not stop them from proliferating. We need to exterminate these beasts, if we could find a way to do so without prejudicing the tractors of burden in neighbouring Asia.

African swine fever is rapidly spreading, it has already spread through China, south east Asia and Indonesia, and is now in Papua New Guinea. If it gets to Australia, as seems inevitable, it could infect the massive feral pig population, which may or may not be a blessing in disguise; it will either wipe them out, or establish a permanent reservoir of the disease. This is a highly contagious infection, more so than foot and mouth disease and much more lethal. If foot and mouth reached Australia and couldn't be contained, quite apart from its effect on domestic sheep, pigs and cattle, it, too, could spread to feral pigs, and also to buffalo, camels, wild cattle, feral goats and deer, which would necessitate a military-type operation to exterminate those animals, if such were possible.

The Fall armyworm has been detected in Australia (Cape York) for the first time. The "worm" is the caterpillar larvae of a moth that has devastated crops around the world. The caterpillars feed, in large numbers, on more than 350 types of grasses and other crops, including rice, wheat, sugar cane, sorghum, cotton, and many fruits and vegetables. This armyworm has the potential to cause major economic damage to agricultural industries. It has already spread right through to Western Australia.

A serious disease of bananas, Panama disease (the TR4 one), has also been imported and is threatening the whole banana industry. Bee populations are now threatened globally by the Varroa mite (Varroa destructor), now imported to Australia, which could have a serious effect on the pollination of native plants and domestic food plants. There is also the African beehive beetle (Aethina tumida), a destructive pest of honeybee colonies, which with their larvae, damage combs, stored honey and pollen and pollute them with their faeces, and may cause bees to abandon their hives. Insecticide residues are another threat to bees.

Not so well known is that Australia now has an imported fungal disease of frogs which occurs world-wide and is decimating frog populations. Frogs are an early indicator of environmental pollution, much the same as the canary in the coal mine. They are important in the ecosystem because they eat mosquitoes and flies, and these could increase greatly in number if the frogs perish, allowing insect-borne diseases to proliferate.

The discharge of ships' ballast water is also a problem because it directly imports marine pests and diseases, we have seen several examples of this, there was a bad one in Darwin. That water should be disinfected before it is discharged. But if foreign registered ships can't even hold onto their cargo of containers, what hope is there that they will do this.

Then there are introduced plant weeds, most of them listed under the Environment Protection and Biodiversity Conservation Act, 1999. Weeds of many sorts invade the land, not only agricultural land but also native grasslands and forests and threaten to overwhelm them in their natural environment. Prickly pear was a serious one, fortunately handled by the introduction of a biological control, Cactoblastis. Another cactus, the extremely spiny

Hudson pear has run rampant in the Lightning Ridge district, now under experimental biological control by the cochineal insect, which is related to Cactoblastis. Prickly Acacia (Acacia nilotica) is a very invasive introduced tree that is spreading rapidly throughout semi-arid Australia. In the south, particularly in the Gwydir River floodplain west of Moree there is an infestation of water hyacinth. In the outback, the introduced Buffel grass is highly invasive; it is spreading like wildfire and easily out-competes excellent native grasses such as Mitchell grass, and in more remote areas is causing a decline in native nitrogen and phosphorous in the soils. Our northern riverbanks are chock full of weeds including bellyache bush (Jatropha gossypifolia) which is spread inexorably by floodwaters, and as well, forests are being colonized by smothering foreign vines such as the American wartime bittervine. Weeds in agricultural land are a major and expensive problem. Amongst them are Parkinsonia, Nagoora and Bathurst burrs, skeleton weed, Parthenium, Hyptus, thistles and many more. These are spread by careless agricultural management, and traffic to and from infested areas.

In the Top End of the Northern Territory in recent decades there has been ecosystem degradation, habitat loss and species decline due to invasion of introduced species including Gamba grass (Andropogon gayanus), mission grass (Cenchrus polystachios, syn. Pennisetum polystachion) and annual mission grass (Cenchrus pedicellatus). The main grass around Darwin used to be annual sorghum (Sorghum stipoideum) and the dry season fires were just low-temperature grass fires, but these invasive grasses that have replaced it now carry very hot and destructive fires which damage and eventually destroy native species. Then there are cobbler's pegs (Bidens pilosa), horehound (Hyptis suavolens) and other virulent herbaceous weeds that also carry devastatingly hot annual fires In the northern wetlands Para grass (Urochloa mutica), olive hymenachne (Hymenachne amplexicaulis) and Eleocharis have invaded wetlands, and there is mimosa (Mimosa pigra) in the floodplains. The now wide-spread floating fern, Salvinia, one of the worst water weeds in Australia, is now across the Kakadu wetlands, with the capacity to double its coverage in two or three days and block out sunlight for submerged species.

Some imported feral animals can be domesticated and put to profitable use, and this may be the best way of dealing with them. For example, feral goats have been rounded up for meat and mohair production, northern buffalo have been captured and domesticated for meat and hides, feral horses (brumbies) were used as mounts in the first world war (termed Walers because a lot came from the early colony of New South Wales) and are still mustered for stock work, feral camels are being put to use to produce milk, meat and hides, and rabbits are there for the taking. Perhaps more use could be made of other feral animals. The issue of dingoes is contentious, with prominent ecologists claiming (with precedent) that the environment needs a top predator and that the dog problem is really due to domestic dogs, which were originally imports, gone wild, and inter-breeding with the dingoes.

One of the best ways to ensure the conservation of native animals may be to allow them to be kept as domestic pets, if conservation institutions and governments would allow it. Pet shops and garden centres would soon accept and accommodate the need for different pet foods and products. A public education program might be necessary to advise on proper foods for each animal. If we could all have a possum or a hairy wombat or a koala or sugar glider for a pet they would be valued and would reproduce like cats and dogs, instead of being isolated in their current institutional situations - "reserves". The old greenie adage about letting them all be free in their native environment is emotive, anachronistic, ignorant and fallacious, and needs to be overturned. This perceived link between the survival of endangered native animals and their "environment" has to be seen as simply farting against thunder. Save the environments, yes, but the incumbent native animals can be protected separately. These native animals' original environments are already destroyed or doomed, despite all those small refuges. For example, we have almost destroyed the habitat of our precious koalas but are not allowed to keep them as pets. Some native animals do require special habitat conditions or are untameable or dangerous and would not be suitable as domestic pets. Some people already sponsor animals from our cyclone ravished habitats, e.g. the Cassowary, a bird you would need to be careful of keeping as a pet anyway. We are allowed to keep imported pigs, cats, dogs and exotic fish as pets, the most destructive pests our environment has ever known, but not harmless native animals.

As an example of the domestication of native animals I once saw a sugar glider living in a ceramic pot on the mantelpiece of a vet who was a friend of mine come out and fly around the house in the evening, going outside even, but coming back in and landing briefly upon the shoulder of its mentor, then it just went back to its bowl. It was completely free but was sheltered and fed as a pet. We have our own dingoes and native cats (quolls) which blend into nature and help keep a balance of numbers between life and death, and both can be domesticated, albeit with care. And, of course, many native birds and fish as well. Zoos and animal refuges are marvellous in the way they tame and virtually domesticate native and so-called wild animals, and so could the public at large.

It is important to distinguish between what is feral and what is native to Australia. If a domestic pet is a native and it escapes into the wild it is likely to resume its original role in the environment and blend in naturally again and do no damage. Compare this with the release or escape of a non-native animal that might become feral, which is tantamount to letting the fox into the chicken coop. Despite those who see Australia as a semi-barren, dry, hard and arid place they are wrong, it is truly a soft and vulnerable country, "fragile" to the Aborigines, and we do not need foreign imports cavorting around doing untold damage as feral animals. Ferals are capable of imposing great destruction upon our land and amongst our native fauna and wetlands, as they have already done, witness the rabbit, the dog, the fox, cats, pigs, goats, carp, camels, donkeys, buffalo, the cane toad, the red fire ant, and the varroa mite as examples.

Associated Issues

Renewable power supplies

E nergy is of paramount importance to civilization globally, but our current means of generating it is producing emissions of carbon dioxide that are causing climate change. We have come a long way from flaring torches, burning wood, candles, oil lamps, lumps of coal, horses and bicycles, to where we now have plentiful industrial coal, liquid fuels, natural gas and electrical power, the latter three being widely reticulated.

However, as a populace, we are currently relying on fossil fuels for most of our energy needs, a fact that is sponsored and protected by large multinational companies seeking on-going wealth. The fossil fuel industry is directly opposed to, and in denial of, any notion of resource depletion or global warming, and will stop at nothing to protect its financial interests. Fossil fuel is now the biggest industry in the world and its proponents are working assiduously to preserve their interests. These include activities such as the exploration, mining, refining, transport, advertising, and retailing of fossil fuels, and support of industries such as electricity production, the motor vehicle industries, transport, production of steel and aluminium, provision of infrastructure, national defence and offence, and business and commerce, all of which involve the profligate consumption of fossil fuel.

Since prehistoric times, civilization has been driven by fire. Fire engenders strength, authority and control, and such power is the manna of the human psyche. It is no longer just burning wood for warmth and cooking food, it is now burning fossil fuels to power industry and society, and to produce explosives and weapons of war. Every aspect of our lives is powered by electricity derived from burning fossil fuels, with a few minor exceptions. Motor vehicles on the roads and air and sea transport rely upon internal combustion engines and turbines, which burn fuel to produce motive power. In each case, serious pollution of the atmosphere and waterways by the down-line products of combustion is the result.

The energy now being produced from fossil fuels is bioenergy that was sequestered and laid down during the Carboniferous Period about 280-360 million years ago when the climate was conducive to the prolific growth of forests, so the energy is essentially photosynthetic in its origin. As we exhume and burn this material, we are reversing the process and releasing that sequestered carbon back into the atmosphere as carbon dioxide gas. This has been happening principally over the last century, or even less.

Coal mining is an extravagant user of water. About 80% of it is freshwater taken from water bodies or groundwater. Most coal mines in Australia are in Queensland and New South Wales. The black coal industry uses enough water to serve five million people, which is more than Greater Sydney uses, and about 30% as much as is used by agriculture. Coal uses 120 times the water used by wind and solar electricity generation to produce the same amount of electricity (2020 figures).

Widespread substitution of sustainable, non-polluting sources of energy is a very important means of ameliorating global warming and climate change. The potential to reduce atmospheric and water pollution, and collaterally improve our quality of life is right before us. We know how to do it, all it takes is action, which, because our political persuasion is so overtly populist, lacking in foresight and ill-informed about scientific and technological developments, will rely upon economics. It will also rely upon removing the strangle-hold of the fossil fuel industries, which may be a greater impediment.

The use of renewable energy sources for electricity production is rapidly emerging as an alternative to the use of fossil fuels. There is much potential in systematically exploited solar power. Enough solar energy falls upon the surface of the earth every day to power all the world's needs for a year, but we harness very little of it. This is similar to our use of other natural energy resources such as wind, tidal, geothermal and hydropower, where we use only a small proportion of what is available. The beauty of electricity is that it can be produced in-situ but is then readily transmittable over any terrain and for large distances, as distinct from water, which requires reservoirs and pipes and may need to be pumped.

The new generation of solar cells is very efficient, and capturing solar radiation has become easy and routine. The major difficulty has been that it can only be captured during the day, and mainly on sunny days at that. Maintenance of baseload power when these sources are less available has been a problem, and development of storage systems is still in its infancy. Research is rapidly gaining ground. Inverters and batteries are becoming available that will convert and store power for long periods. Storage batteries used to be lead-acid types, then lithium ion ones took over because lithium is the lightest natural metal in the world and carries an electric current very efficiently. The advent of electric cars induced research into safer, smaller, lighter, and more powerful batteries, some of them using liquid or gel electrolytes and solid-state technology. These could hold twice the energy of conventional lithium ion batteries and offer twice the range in electric vehicles, and may need only minutes of time to recharge. A major research step is the development of zinc bromine gel batteries. These are cheaper than lithium ion batteries because the core materials are abundant and inexpensive, are long lasting, not a fire risk, and can be made portable. These new batteries are still in a developmental stage but some of them have been adapted for and are already in experimental use in motor vehicles and domestic homes. Battery technology has been accelerated by advances in android mobile and iPhone development. However, the problems of rectification, storage and transport of bulk solar power have not yet been adequately resolved, and alternative distribution points for refuelling of vehicles will take time to provide. Maybe one day electric cars will be equipped with solar panels on their roofs.

Domestically, the new batteries allow a day's solar energy capture to be stored and come on stream during the night thus providing a 24-hour supply, even sufficient reserves to withstand a whole cloudy week, because some solar energy does come through clouds. Widespread use could contribute to the national grid and allow peaks and troughs in supply to be alleviated. The prospects now are for most homes and housing estates, even whole suburbs, to be powered by renewable energy, doing away with coal and wires. Sunny outback properties would then be independent of the grid or diesel fuel, although most would still need a portable generator for such things as welding; heavy industry such as iron and aluminium production likewise. With improving technology and more widespread application costs are coming down as well. The next step, which is under review, is to improve the portability of the new storage units so that they can be used in vehicles, boats and perhaps even aircraft. There are already experimental electrically powered aircraft in flight.

Under conditions of extreme heat when solar radiation levels are very high and conventional grid power supplies are stretched to the limit and may become unable to cope, it would seem logical that solar power be called upon to contribute. It can also be used to pump water to a higher potential energy condition for storage, to be used later, perhaps at night, to develop hydropower.

Wind power is now cheap and easy to produce and already contributes significantly to the national grid. Electricity generation depends on the occurrence of wind, which is more reliable in some areas than others, but whenever wind occurs whether during the day or night it can be utilised. Wind turbines are sometimes denigrated but on completely spurious grounds; they are aesthetically elegant, and a visual example of beneficial modern technology. Noise is not an issue; some people complain about even the twittering of birds. Some kinds of wind power are unbelievably strong, such as the immense power in hurricanes, typhoons and cyclones, if we could but tap into them.

Hydropower is an excellent and non-polluting energy source, driven essentially by gravity. The Snowy Mountains Scheme is a well-known example; there could easily be other similar schemes, there are suitable sites in Tasmania. It is estimated that there are thousands of potential but untapped sources of hydropower in Australia, many of them quite small but substantial and reliable enough to contribute, even locally. Wherever there is flowing water there could be hydropower, and the outflow can still be used for other down-line purposes. A huge benefit of hydropower is that it does not require storage batteries, it provides energy all day every day without interruption.

There is tremendous potential for additional hydropower generation in northern Australia where there are some very large rivers. In full flood the Victoria River east of Timber Creek in the Northern Territory fairly erupts, it is about a kilometre wide and reckoned to be very deep; it has several times over-topped the bridge, which is high above its bed. As it sweeps down towards the sea it carries as much water as the Daly, the Adelaide and a couple of other rivers put together. Then there is the Ord near Kununurra, another very big river, with a full supply level in

Lake Argyle of 18 times Sydney Harbour. If a one in 100-year rainfall event hits the Ord catchment and fills the dam to its maximum it could hold up to 70 Sydharbs and would overflow the spillway for more than a decade, with large standing waves upon the torrent. Full of crocodiles, too. Intrepid individuals surf upon the spillway when the dam is at high levels. In the north of Western Australia, the Fitzroy is another big one, one of the largest in the world. In high flood it is said to release to the ocean enough water in just three hours to supply Perth for a year, or to fill Sydney Harbour in six hours. The Burdekin is also huge. In full flood flow it is said to run enough water out to sea to fill Sydney Harbour in only five hours. Of these, only the Ord and the Burdekin are dammed, but apart from a small scheme upon the Ord, and a proposal for a larger one on the Burdekin, they do not yet produce much hydropower.

There is a proposal, in a modified version of the original Bradfield Scheme, not yet fully researched, to build a very large dam on the Hells Gate River in coastal north Queensland which would divert some of the flow from the South Johnstone, Tully, Herbert and Burdekin Rivers, which would otherwise flow into the sea, inland through tunnels, under the influence of gravity. This scheme, were it to be found feasible, is said to be capable of generating large amounts of hydropower and to provide enough water inland to command an irrigation area bigger than Tasmania. There are other versions of this plan. A feasibility study would need to include a scientific assessment of the effect of that water on those soils in an irrigation environment; the Ord Scheme had problems of this sort. Unless irrigation was found to be a viable proposition, such a scheme may be better employed in replenishing the Murray-Darling system, where water for irrigation and domestic use is in short supply and suitable infrastructure is already in place.

Tides are driven by the gravitational pull of the sun and the moon and are entirely predictable. Power can be generated by their rise and fall, and this power is everlasting, renewable and completely non-polluting. Tidal power is being developed in England where large coastal lagoons are built that fill with sea water on in-coming tides and release it again as the tide falls, turning strategically placed turbines both ways. Several other countries are also undertaking serious research. This works best in areas with a consistent, high tidal range; it is not so good at neap tides. Tidal power generation is costly but could be a significant contributor to the provision of renewable energy. As well as tides, it has been speculated that if only 0.1% of the energy in the world's ocean waves could be tapped, it would provide enough energy to satisfy the entire global energy needs five times over! Wave energy is of course largely driven by wind.

And why not geothermal power? Power from down in the super-hot crust of the earth? It is just beneath our feet; the technology is well known and already in use – ask the New Zealanders. It is unlimited, everlasting, operates ceaselessly all day and night, and is completely non-polluting. Even existing artesian bores could be adapted for outback station or community use; some of them have near boiling water heated by geothermal energy, such as at Birdsville where it is already in use to generate electricity, and there are plans for expansion. For some applications we do not need super high temperatures, for example to make a heat pump that will operate an air conditioner all we need is a temperature differential, which is available at shallow depth.

Hydrogen is the most abundant element in the universe and is a basic constituent of water (H_2O). Two thirds of the surface of the earth is covered in water, so there is an inexhaustible source everywhere at hand. Hydrogen burns extremely well, and upon combustion (combination with oxygen) the product is only pure water, so it is completely non-polluting. Hydrogen can be isolated from water by electrolysis, but the process requires an input of energy. That energy can come from renewables like solar, wind, etc. or from the burning of coal or natural gas. In the longer-term, generation from renewables would be ideal, but in the transition, fossil fuels may need to be used. Research into the solar electrolysis of water to produce hydrogen is continuing and promises to be a large industry in the years ahead. There is a proposal in Queensland to produce liquid hydrogen using Australia's abundant solar energy resources and export it, negotiations have begun with Japan. Hydrogen can be used to power industry and domestic homes directly, much as natural gas does today. Fuel cells that produce hydrogen are already in use in stationary modes and to power motor vehicles. Production of hydrogen by electrolysis of saline water in "batteries" is already in use in portable camping lights.

Biogas is a mixture of gasses produced by the breakdown of organic matter in an anaerobic digester by a process of fermentation. It can be produced from raw materials such as animal manures, especially pigs and cattle, agricultural and horticultural residues, municipal sewage and urban food waste, most of which goes to waste or into landfill. It consists primarily of methane and carbon dioxide, but the CO_2 can be removed, and the methane compressed and used for purposes such as space heating, cooking, or to power motor vehicles, similar to natural gas. The process generates no new carbon dioxide as the CO_2 was originally absorbed from the atmosphere in the growth of the primary bioresource, it is simply recycled. Digestate, the material remaining after the digestion process, contains all the recycled nutrients that were present in the original organic material but in a form more readily available for plants, and can be used as an organic, non-chemical fertilizer. Biogas can provide a clean, renewable, and reliable source of baseload power in place of coal or natural gas, and provides an alternative route for waste treatment.

Then there is nuclear power, a much debated and often disliked energy source because it generates potentially polluting radioactive waste, and there is the danger of a power station meltdown. It is also possible to generate fissile material in a nuclear reactor that could be used to produce atomic bombs, which could be dangerous in the wrong hands. Nuclear fission involves the splitting of atoms of uranium or related radioactive materials to generate heat and steam, which is harnessed to produce electricity. There are many nuclear power stations operating around the world including desalination plants and they are now much safer than they used to be. Nuclear power plants do not burn fossil fuel and do not produce greenhouse gas emissions. Australia produces one third of the world's uranium, most of which is exported. Research today is into nuclear fusion, a different process, which could produce even more energy, but so far plausible schemes to harness it are in their infancy.

Distillation of water from the air is now a reality. In one application, solar panels and hygroscopic technology are used to condense water from atmospheric humidity, providing pure water that is free of chemical and biological pollutants. Commercial units are free-standing and require no external electrical connection. Current models produce several litres of pure water per day and can function in even very dry climates. They are also zero-waste, because the smaller units are portable and operate on-site so there is no need for distribution of the water, so no plastic bottles. In arid countries, solar hydropanels can provide a renewable, infrastructure-free drinking water solution that can be used in the driest countries on earth, such as Australia, and also those where water pollution is a major cause of sickness and death. Another example is in simply saving the condensed water from de-humidifying equipment in common use such as air conditioners.

There are other interesting research developments. One is genetically engineered bioluminescence, produced by a common bacterium with introduced genes from luminescent squid, which produce light that can locally be as strong as daylight.

Renewable energy sources are everlasting, inexhaustible, freely available, do not have to be mined, and the development technology is a lot simpler than trying to contain and sequester emitted carbon dioxide, and cope with all the downstream pollution and health issues of fossil fuels. Once it is more widely accepted it will also be a whole lot cheaper. With renewables, people will be able to tailor their needs and costs to suit themselves, rather than relying on a fossil fuel-based energy grid. This is slowly happening around the world with community-based energy projects, 70 or so in Australia alone, based largely on solar and wind power.

Given further technological development, if it could be made financially competitive, as it assuredly will, and if the fossil fuel industries were to allow it, a combination of renewable power sources could easily cover all our needs. Unfortunately, the move towards renewables so far is not displacing the consumption of fossil fuels because total world energy demand is increasing at a more rapid rate; renewables are not substituting for fossil fuel use, and fossil fuel use is still increasing.

Our thinking on the generation of power is out of date. Now, in order to go forward, we need unambiguous, future-looking planning that will deliver predictability and stability, to provide an environment where industry can confidently develop a new approach to energy generation. First, and most importantly, we must move to a zero emissions source of electricity generation. Electricity generation from coal is now clearly uneconomical, renewables are increasingly becoming cheaper and are smashing the economics of coal-fired power stations. Renewables plus storage is where we need to go. This cannot be done by decree; people must vote for governments that will provide better political will. At present, this is being blocked by vested interests and the weak subservience of current politics. As an example, the move to renewables is clearly evident in Germany, where gross power generation from renewables is escalating steeply, and that from anthracite (black coal), lignite (brown coal), and nuclear fission is declining. Consumption of natural gas and oil remain about the same. In the United States, renewables are forecast to produce more electricity than coal for the first time in 2020.

Treating the symptoms, not the cause

One of the failings of the modern world is a tendency to see problems at face value and to proceed with solutions to them by addressing the issues directly. This is illogical, and generally will not solve the problem. Solutions rarely lie where the effects are and addressing the symptoms will provide temporary respite at best. Problems should always be addressed at their source because cause always precedes effect. We see this failure commonly in situations where man is in opposition to nature, but also in business, socioeconomics, and politics. Man's concerns are highlighted, but the causal environment is commonly ignored.

The occurrence of cyclone Debbie on the Queensland coast (2017) and its aftermath provides a good example. There have been six major floods in Rockhampton in the last decade, leading amongst other things to much higher flood insurance premiums. After cyclone Debbie there was a call for higher levees along the Fitzroy River banks to protect the town from floodwaters. However, these calls were from ill-informed sources, because to do so would be treating the symptoms, the flood itself, not the cause of it. The primary cause of the flooding lay in the catchment of the river, where long-term unwise land use practices had reduced the infiltration of rainwater and increased the amount of peak runoff, which shortened the time taken for the floodwaters to reach the city, resulting in greater flooding. With levees, the floodwaters would be constrained within a channel rather than being able to spread out across the floodplain, river heights would rise accordingly, and water velocities would increase. There would then be downstream effects, where the flooding would be passed on to residents further down the river.

However, in contrast, as the remnants of cyclone Debbie progressed down the east coast, the Bureau of Meteorology gave clear information on the cause of subsequent events. As Debbie moved inland it lost its eye and became a rain depression. As it drifted south it was met by a terrestrial cold front from the west which interacted with it, and between the two of them they produced heavy rain and severe storms along the coasts of Queensland and New South Wales. The original marine system had been complicated by an on-land weather event. High tides made it worse near the coast. This Bureau of Meteorology report was a good example of looking at the cause, the process, rather than just the events themselves in isolation.

A similar example is interference with fluvial systems, as regional Councils often do, by cutting off meanders in rivers to facilitate disposal of floodwater from flood prone urban areas. That action will certainly solve the immediate local problem, but because of the increased channel velocity erosion of river-banks will occur, and the flooding will be passed on at greater severity to downstream locations. The cause, however, lies in the hydraulic characteristics of the stream. River meanders are a natural feature of drainage systems. In lower tracts they increase the length of the drainage channel and so reduce its gradient and the velocity of flow. Sediment settles out on the adjacent plains as a result. Periodic inundation of floodplains is a natural event and ought to be accepted as a recurrent phenomenon, and if it is contrary to the needs of urban communities then these areas should not be so occupied.

Recent mudslides in Bolivia and the Philippines were met by declarations that channels and barriers would be erected to prevent any future such events from impinging on the towns affected. However, that would be to treat the symptoms. The mudslides were caused by the clearing of land in the surrounding hills which reduced the removal of water deep within the soil by the trees, then cultivation, which increased the surface infiltration of rainwater, and the planting of corn, a shallow rooted crop. Excess water was not removed by the crop and was able to accumulate deep within the soil, until positive pore pressure caused the mudslides to occur. A relatively small earthquake is often the trigger. Whole hillsides can come sliding down and engulf local villages. Devastating mudslides are not limited to the under-developed world, as the recent (2018) calamitous mudslide in southern California shows. In that case the cause was more complex because intense fires had destroyed the vegetation over large areas, not clearing, and the very heavy rain that followed saturated the soil and precipitated the mudslide.

Business and commerce treat the symptoms for a very clear reason. If a symptom such as obesity (for example) persists, money can continue to be made by the food, diet and pharmaceutical industries by continuing to administer to it. If obesity was to be contained there would no longer be profits in it, and for them, that would be killing the goose that laid the golden egg. Liposuction and gastric banding are treating the symptoms of obesity, not the cause, which is essentially dietary. The food industries are careful to maintain and even extend the problem, they are not silly, they know very well what makes people fat, they just want to continue getting their profits. Subsidizing dental treatment rather than dealing with its cause, which is poor diet and oral hygiene, is another example.

Then there is the disposal of domestic waste. To send material such as plastic to landfill is to treat the symptom, the end-point problem, rather than going to its cause. The cause is the production and use of this plastic waste in the first place. The cause of plastic pollution has been food-producing and marketing companies packaging food in easy to handle plastic containers and bags making shopping more convenient and attractive for members of the public, and so attract more sales.

The recent decision to replace single use plastic bags with re-useable ones has its downside. The old bags were made from high density polyethylene, and the new ones are mostly made of polypropylene, another plastic. Both may be biodegradable, which means they break down completely, or just degradable, so that they break down into small pieces which can be ingested by aquatic animals, or they may be completely inert. From an energy point of view the new bags need to be re-used multiple times because one new bag uses 28 times the energy to produce than one old thin plastic bag. Bags made of paper, cotton, canvas and linen cost a lot more to produce in terms of energy and water. The old single use bags were commonly re-used as bin liners, but commercial bin liners are the same as single use plastic bags and are used only once. A problem with the new shopping bags is that they accumulate fungi and bacteria that can cause infection, and they need to be washed regularly. Levels of coliform bacteria in them have been found to reach 300 times the safe level in drinking water. Keeping items such as meat, fish and poultry in a separate bag that is regularly washed is a good idea.

Recycling treated sewage water for domestic consumption is another example. It treats the symptom, which is a shortage of drinking water, but fails to acknowledge the cause, which is flushing our toilets with clean drinking water when grey water would be perfectly adequate. The recycling should be done at the start of the cycle, in the home.

Perhaps a sad one concerns the worldwide demise of coral reefs. The Australian government is intending to spend millions of dollars to mitigate the effects of starfish predation, chemical pollution and sediment influx on the Great Barrier Reef to protect the coral, and also the thousands of jobs in the tourism and hospitality industries. This is treating the symptoms, because the cause of coral demise is mainly rising sea temperatures due to global warming and associated coral bleaching. Sea level rise may also be becoming a cause, a new one in the pipeline.

Commercial podiatrists specialise in providing corrective footwear for people with painful or malfunctioning feet. I visited one once when I had an Achilles heel injury and was offered footwear that would virtually immobilize my feet, the idea being to treat the symptom by cosseting the injury. I demurred, and visited a sports medicine doctor instead. He offered quite the opposite treatment; he told me that immobilizing the feet would only make them weaker and more vulnerable, and that I needed to provide stimulus to the feet which would assist them to strengthen themselves and repair the damage. He advised me to take off my shoes and socks and go barefoot, over as rough a surface as I could stand. He was treating the cause of my injury, which was over-cosseting of my feet, and it worked.

Politicians are also prone to treat the symptoms, not the cause. Politicians do not lead, they follow. During election campaigns they promise the public all the things they can think of that will win votes, and those things are always what the public want immediately. However, what the public want is invariably relief from symptoms, things like unemployment, housing affordability, taxation, the cost of fuel, electricity, water, education, food, etc. but many voters are not aware of the need to deal with the underlying cause of these problems, or that by being given token "solutions" they are just making their plight worse in the longer term. Contributions of government relief measures and subsidies such as the first home buyers grant always finish up in the hands of the vendors, and result in even higher prices. House prices will only drop if demand does, so subsidising demand is counterproductive. Only statesman-like politicians will advocate measures that go to the cause of problems, but they usually don't last long in office; they are either replaced in their own party by another more symptom-ori-entated member or booted out at the next election. The welfare of humans is increasingly dependent on longer than immediate term views, i.e. the causes of these problems, but that doesn't win votes.

The medical profession is well aware of the symptom/cause situation and where possible, doctors treat the cause of any bodily malfunction. But there are times when the cause is not clear, and they must resort to treating the symptoms until the cause becomes apparent. Painkillers are a good example. Placebos work by suggestion, where the patient is purposely led to believe that a particular treatment will relieve or cure the problem, and they often do, but they have no clinical effect. This is a bit like the aboriginal "pointing the bone" but designed to have a beneficial effect rather than a lethal one. By treating symptoms such as those of pain, placebos circumvent the need to discover the cause of the affliction, which may be referred pain or a psychosomatic illness anyway.

The habit of preaching to the converted is similar to that of treating the symptoms. The road toll is a case in point. Regulatory authorities repeatedly provide clear messages about road safety, but these are (unwittingly, it seems) delivered to those who watch rational television broadcasts and internet sources or read normal newspapers and so become aware of these messages, but these are the converted, who don't so often transgress. The main perpetrators of road trauma are the young and the intoxicated, who don't observe those media sources. Communication of information on the road toll and the effect of drugs and alcohol on it requires saturation coverage in the social media, not conventional newspapers and television, to contact the chief offenders. Treatment of the cause is not being delivered to where the symptoms are occurring. Increased regulatory control is only part of the answer, education of all road users, new and old, and its effective implementation are imperative.

Societal dysfunction is one of today's widespread problems. There are many forms; domestic violence, abuse of alcohol and drugs, crime, bullying, disrespect for authority, and social media misuse to name a few. These tend to pass on from one generation to the next and become replicators. The symptoms are clear enough, but what are their causes? Some think that increased population density, the seemingly insoluble mega-problems of the world, and dereliction of duty by parents are probable contributors. Whatever they are, increasing regulation and punishment for transgression will not help much; no solution will be found until the causes of these dysfunctions are discovered and treated at source; but that could be very politically incorrect.

There is much concern about poverty and hunger in such countries as Yemen, Myanmar and much of Africa, a situation that is very disturbing. But the people there keep on making more and more babies! It is pointless helping because the people continue to reproduce prolifically, and until they stop doing so we are merely treating the symptoms, not the cause. Saving the existing children is of course necessary, but why not treat the cause of this at its source, and stop making more and more babies? The basic cause is the unbridled impulsion of humans to procreate. It seems that the very last thing that distressed humans will do is stop making more babies. A tragic and seemingly insoluble case of treating the symptom and totally ignoring the cause.

After a terrorist inspired explosion, crowds of people stampede and kill or injure more of each other in their rush to get away, but by then the time of danger has passed, the bomb has already gone off and conditions afterwards are much safer than they were before the blast. Why panic, when it's all over? Similar events occur with foreign nationals after a terrorist attack in the country they are visiting, they leave the country in droves and go home, when to stay is then much safer than it was before. This response ignores the cause, the attack, which is already over.

Many events are of course beyond human control. Many countries are beset by famine. The causes of famine are multiple, and include weather, politics, corruption, and overpopulation. In many cases these causes cannot easily be addressed and the only option is a palliative one, to treat the symptoms by provision of food, medical aid and shelter, sourced from outside the affected areas. The effects of volcanic eruptions, earthquakes, tsunamis and collisions of extra-terrestrial objects with earth are beyond the scope of the residents of earth to mitigate, and cause and effect become coincident. Severe weather events such as calamitous storms, droughts and floods may be rendered more severe by the effects of global warming; global warming is a tricky one, we seem to be on the cusp of averting a catastrophe, but are we? There are still those who want to simply treat the symptoms.

Top-down versus bottom-up

These are two types of approach common in the administrative activities of mankind and it is useful to be aware of them. Activities funded by government espouse two main methodologies, top-down and bottom-up, usually both at the same time, and they are often in conflict. The most common of these occur in overseas aid programs, and politics. This is important because both kinds of administration affect populations of all sorts, the donors and the recipients, worldwide.

International aid is provided by consulting companies in the Western world financed by national funding agencies, with the aim of raising standards of living and reducing poverty and disease in societies unable to progress by themselves. Agencies charged with delivering and implementing foreign aid programs usually provide multi-disciplinary teams of consultants. All are specialists in their various fields and are technology literate, internet savvy, and widely experienced in the delivery of aid around the world. The usual scenario is that the recipients of aid aren't broadly informed, they know what they want, but may not know how to get it or fund it. These people are largely parochial, lack education, may be semi-literate, and have difficulty envisioning what the solution to their problem might be. They know the symptoms they are experiencing very well, are aware of their historical, cultural and social constraints, but they are often not clear about the cause of their problem, which is an essential starting point in finding a solution, and so they don't know how to achieve a satisfactory outcome.

"Top-down" means that the program of aid is determined by the more biophysically or academically orientated specialists, who are largely the scientists. The top-down people know what the options are, how to provide them and what off-site or other effects there might be. "Bottom-up" means that the nature of the delivery is determined by advocates of the recipients such as sociologists and anthropologists, who sympathise with and work directly with them. The balance of these approaches and the most appropriate methodology is loosely defined in Terms of Reference, which are sanctioned by the donor agency.

The sociologists are usually much in sympathy with the local points of view and usually speak the local language well. In turn, the biophysical scientists are often less aware of the nuances of custom and culture and probably don't speak the language quite as well. However, neither approach can work on its own, both are required and need to be integrated, and here is where the dispute is at its most vituperative. For a successful outcome, each faction really needs to depend upon and cooperate with the other.

An example is a project in Indonesia where the objective was to stop the environmentally damaging "slash and burn" shifting agriculture carried out on steep hillslopes. Traditionally, these areas were cultivated for two or three years at a time in a rotation of locations that provided several years of fallow, which allowed the native scrub and bamboo to regrow. The hillslopes were favoured because periodic burning of the regrowth not only cleared the area but provided ash as fertilizer for the coming crop, and also because the soils there were softer and easier to dig. The issue was that the fallow period was reducing, the soils were becoming overused and increasingly infertile, and soil erosion was becoming a serious problem. Large areas were affected. The project principals wanted to make the growing of annual crops of corn and upland rice more permanent by transferring them to the alluvial flats in the valley floors, thus overcoming the erosion problem. Appreciating the difficulty of cultivating these much firmer soils the project undertook to bring in tractors and cultivate the lowland soils, to enable relatively easy planting of crops. From their "bottom-up" position the locals said that it would never work. What was discovered when the initiative failed was that when the dense grass of the lowland soils was turned in by the cultivation and rotted, it released oleophobic compounds which impeded the growth of any grass for two or more years. Corn and rice are both grasses. The locals were unaware of the scientific reasoning, but they knew quite well that this would happen. They needed a reliable crop every year, they could not wait for two or more years, they would starve. In this case top-down seemed logical, but bottom-up proved to be the more realistic.

Despite any dispute between project protagonists, upon completion, the implementation of any international project is totally reliant on the recipients actually carrying out the findings. One cannot impose a regimen upon a community except under a despotic tyrannical dictatorship, which is all top down anyway, because if the people decide they don't approve of it they won't adopt it, and so the project won't be implemented and will have been a waste of resources. A good approach may be to embed all aspects of international project implementation into existing structures, both community and administrative, so that life goes on as usual with the same responsibilities, but additional or alternative terms of reference apply.

There is an analogue in politics. There are many issues in politics but amongst them are issues of a top-down or bottom-up nature. Of Australia's two major political parties the more liberal persuasion takes a longer-term view, and the more trade union orientated one takes a shorter-term and essentially "populist" stance. The voting public is about equally divided on which is the better course and so governments regularly change, and when they do each tries hard to discredit the preceding one for the perceived mess that they created and tries to reverse the direction that was taken; the parties are in perpetual conflict. To a politician "fix a problem" means "get elected"; they have been well labelled "retail politicians".

153

Politics revolves around the catchword "jobs". The "longer-term" politicians argue that the immediate symptoms of social distress will only abate if the longer-term causes of it are dealt with; they focus on policies designed to improve the employment environment and reduce the cost of living at source, and argue that to support small business, commerce and industry, which are by far the biggest employers, is in the longer term to guarantee the creation of more jobs and national wealth. They also promote the maintenance of a budget surplus and the importance of national security. This could be seen as a "top-down" approach by government because the politicians are cognizant of cause and effect and are trying to lead by addressing the cause. These politicians are supported by a more thoughtful and educated section of the public.

The "bottom-up" politicians are supported by those who want their current needs dealt with immediately, even if only temporarily and superficially. These people do not generally take a longer term view, they have less awareness of the outcome of different policies and are more concerned about issues such as unemployment, the cost of living in all its forms, and how to make ends meet now, rather than in the future. Bottom up politicians do not plan as much for the longer-term national economy or international affairs, and usually operate by throwing borrowed money at the popular issues that got them elected. This is treating the symptoms of public distress rather than their causes. It could be seen as "bottom-up" politics because the public are leading the politicians. However, it usually results in increases in some forms of taxation, a blowout of the deficit in order to fund election promises, an increase in illegal immigration (the boat people), and a diminution of national security and economic reputation. A significant number of these politicians have graduated from trade union leadership positions, which are notoriously member orientated. The views of the more liberal persuasion are seen as elitist by trade union exponents, who claim that they provide preferential treatment of the already rich and label their opponents as being "out of touch" with the common man. The worse the immediate problems of the public become, the more adherents to this latter approach there usually are.

It needs to be kept in mind that democracy has been defined as "government of the people for the people by the people" (Abraham Lincoln) which, taken literally, is very clearly bottom-up. Fortunately, government is not solely in the hands of politicians, most government is undertaken by government Departments which have a more consistent and experienced staff, rather than directly by political figureheads. The outcome has been labelled as getting the government we deserve.

Technology and unemployment

It has been forecast that within the next decade or so unemployment will become the world's major social problem. We have entered an age of data, or information manipulation. People don't seem to have grasped that technology has already begun to outmode labour and will before long significantly replace it. A rapidly increasing population seeking work is exacerbating the unemployment problem. It is fast becoming a better proposition for business and industry to turn to technology rather than to employ labour, especially in the really big employing companies such as manufacturing, factories with assembly lines, large corporations and multi-nationals, financial institutions, even the public service.

We are now way past the age of the computer, which has already been credited with displacing people from the work force. There are smartphones now which are the new mobile computer, smart TVs, and even smart wrist watches ("smart" meaning they will spy on you). We now have a pace of change unprecedented in human history, driven by technology, and are facing a great wave of artificial intelligence and automation that is sweeping the world. Persons are being replaced by their digital analogues. This is not only so in the more advanced countries; India and China, two countries that hold most of the world's population and have been amongst the poorest, are becoming technology-savvy at a pace that exceeds that of the Western world. The new age of much broader and more sophisticated technology will have all-encompassing effects, to the detriment of "workers" everywhere.

"Jobs" has become a significant election issue. The cost of labour is the big issue here; jobs are being exported to countries that have cheaper labour, like China, India and south east Asia. As just one example, to process one beast at a meatworks in Australia costs up to $400, but only about $50 in Indonesia (2018), and much of the difference is due to the costs of labour. Hence the live cattle export trade flourishes, at the expense of Australian employment. Jobs that do require a human will be fine, such as in the trades, where people need things built or fixed in their homes or workplace that can't be done remotely, by machine, or by overseas workers, and there is actually a shortage of labour in this area. The age of the contractor is here; people with top skills who can choose their jobs, often in a "squad", and dictate their own terms. That is not employment.

An over-arching aspect is that the whole social ethos of "jobs" has changed. In the times of our parents, work was seen as participation in the broader task of nation building, and every able-bodied person willingly joined in and participated. Money was not the only driver; it was a social responsibility and a pleasure to get stuck into doing something for the common good. Workers helped and supported each other, and effort and striving were part of the common weal. Willingness and self-motivation were the primary drivers, and "bosses" were accepted as just one of the workers, with perhaps better leadership qualities or a broader overview. Not so today. Now, a job is seen as a means of personal survival, getting enough money to live on, however or by whatever means it is done. Willingness and self-motivation have been replaced by having to get the necessities of life, and cooperation has given way to competition. Rather than help and support for each other, it is now may the best man win. Bosses are now remote impersonal company directors. Overall, the primary social imperative of "work" has become a drudgery, to be cheated and escaped from as much as possible.

With the advent of artificial intelligence, modern electronic and robotic control of operations is increasing rapidly. Electronics and robots are more reliable, more efficient, absolutely predictable, they work all day and night without a break, are better at their job than humans, have no problems with race, religion, gender, harassment, age, or looks, don't need training or refresher courses, don't take sick or maternity or compassionate or holiday leave or want workers' compensation, are never absent or late getting to work and they never go on strike, cannot sue, and there are no workplace health and safety controls. They require no pay so there is no payroll tax, no promotions of rank, no bonuses, no superannuation, no retirement packages or severance pay, and none of them belong to a Union. If they break down, which is unusual, they simply need to be repaired, then they are back on the job. Humans need to be job-interviewed, are expensive to pay, are subject to government regulations, Union demands and enforcement, litigation, excessive paperwork, do less reliable and often worse work, are slower, less controllable, get sick, need office space, transport, food and water, upkeep of morale, and are subject to human error.

The well-intentioned Union movement is on its way to outgrowing its usefulness and becoming an anachronistic institution. The issue for workers is not better pay and conditions at all, it is rapidly and increasing becoming unemployment due to technology. Now, whenever a farmer or a businessperson contemplates a change or an upgrade to their business, they check out the latest technology before the availability of labour. There is already a strong move towards temporary or part-time work. Installation of robotics is costly but has only to be done once; it is regarded as capital expenditure and is tax deductable. The labour Unions, especially those with compulsory membership (and political parties affiliated with them), are rapidly killing the only goose they know with all its golden eggs, because business and industry are simply going around them. One only has to look around the world to see this happening. The finance sector is rapidly turning to technology and shedding jobs, banks are a good example, automatic teller machines were just the start of it. Increasing numbers of businesses now operate only online, letters are no longer answered, it is all by email, internet and credit cards (cashless), and increasingly via the social media. Even currency is going digital. Today, in any significant transaction, computer literacy and internet access are assumed.

Workers, and their preferred political party endorse an ideology that at least tacitly supports the extractive industries such as mining and logging which rely upon significant inputs of labour, and the labour-intensive industries such as those that employ factory assembly lines. In the former case, widespread concern about biosphere decline and global warming, and in the latter, the increasing application of automation and robotics and the universal adoption of renewable energy, which are more high-tech than labour, will militate against employment.

There are already fully automated factories and assembly lines; driverless cars, trucks and trains; mechanised agricultural operations; satellite-guided tractors; and a whole plethora of drones and robots. Supermarket check-outs have a self-help option that cuts down on staff, and perceptive, completely electronic checkouts are coming. Even the ubiquitous barcode is being replaced by a much faster and more capable "matrix" code, the QR or Quick Response barcode. Look at modern passenger aircraft, they have been virtually flying themselves for years now. When did you last see a human lift operator? Or a petrol bowser attendant? Or a train or tram conductor? Or talk to a human operator on the telephone? You can't get anywhere online without a mobile phone number to be texted in reply. I was watching a grader driver levelling a paddock; he just set up his GPS base-station and let the machine drive itself, no assistants required, just him and his connection to a satellite. Robotic dairy farms are a reality, the cows are milked by robots. Robots can even carry out surgery. Vegetable production is increasingly in artificial environments such that a whole production system can be managed by one person at a control panel. Even fruits such as mangoes can now be harvested by machinery. Access to orbital space is changing too, computers can now fly space rockets better than man can. To cope with the huge weed infestation problem in broadacre farming there are now drones that locate particular weeds and automatically relay their location to a mobile on-ground robotic spray unit, and large paddocks are weeded completely by remote.

The State of Victoria has now lost almost all its motor car manufacturing industry, including Australia's own car, the Holden, and Adelaide its Harley Davidson factory, said to be due to "economic factors"; but what it really means is moving operations to countries with cheaper labour, or re-jigging the factories to be much more completely robot operated. Train drivers of the Victorian Rail, Tram and Bus Industry Union which is affiliated with the ACTU and the Labor Party, are now faced with the reality of driverless passenger trains. Modern mines are being substantially automated, with electronic control panels and driverless trucks. It's becoming a common story throughout industry.

With radically higher unemployment people will find it difficult to support their current lifestyle. Housing will become increasingly unaffordable, not only due to the price of housing, but also to the inability of buyers to raise the cash for a deposit. People who have houses will have difficulty in repaying their mortgage, send their children to university, purchase a car, be able to afford the increasing costs of water, fuel, electricity, rates, licence fees and even food. Insurance premiums will be unaffordable and a thing of the past. Labor/Labour/Republican types of government are preferred by many people (half the population) because they are seen as addressing these problems, but that will only result in the issue becoming more entrenched and will increasingly militate against the employment of labour. This issue needs to be addressed at source, and the basis of it is that employers are preferring technology to employment. There will be knock-on effects also. Governments will receive less tax revenue because the unemployed don't pay tax, nor do robots. More people will need welfare payments. There will be widespread social unrest. Crime rates will soar.

New technology won't affect all jobs, but the nature of employment will change. There won't necessarily be fewer jobs, just different ones. To belong to a trade Union will become a decided disadvantage. Those who have a trade qualification or a university degree in a relevant subject (with good results) will be better off, but only if they are technology savvy as well. Many are already capitulating with the inevitable and working from home or using the internet, online transactions are booming. There is no future in continuing to do things as they have always been done, but there may well be great employment prospects in peripheral and support activities for the new technology.

The digital revolution will initially affect industries and businesses where employees undertake highly repetitive tasks. Small businesses will continue to employ large numbers of people, there will always be a need for social workers/carers, medical practitioners (but online diagnosis and prescription is increasing), police, ministers of religion, media and sales-people, hospitality and construction workers, seasonal agricultural labour, those who pander to the flesh and others. In the immediate future, the jobs market will be very dependent on the more efficient use of resources such as coal, gas and water, but ultimately our reliance on fossil fuels will be much reduced in favour of renewables, and dependence on the energy grid will reduce with it, another blow for employment. In that instance, with the rapid emergence of the renewable energy sector it may seem that one option is to bone up on that technology and seek employment there, or in other fields where rapid change is occurring, for example, digital technology, automation, global warming, energy efficiency, dissemination of information, or dealing with the degradation and pollution of everything.

Science, technology and innovation hold the jobs of the future, and are integral to the benefit of us all. Workers will have to accommodate change, because at the moment, led by their Unions, they are demanding and pricing themselves out of a job. The more workers demonstrate and strike for better pay and conditions the more they will accelerate the trend by business and industry to avoid the employment of labour. Those who can, will adapt; the rest will go on unemployment benefits. Employment is becoming trouble from any perspective, but technology is pragmatic and benign.

A significant developing problem is that technology seems to have exceeded the capacity of the human psyche at large to cope. To be precipitated from hunter-gatherer to technology-savvy man so quickly may be too much, too soon, even for homo sapiens, the most illustrious creature ever to walk the earth. In one generation, from never having been computer literate, and some still not so today, to being exploited by those who are, especially those in commerce and industry, is the lot of those who are unable to compete. These people are largely the "workers".

A Case Study:
The Australian Aborigine

The historical invasion and disenfranchisement of original native populations in newly discovered territories by British and European colonialism can be added to the list of world mega-problems. This began centuries ago, well before the Anthropocene, but the effects have been ongoing and are still very evident. The Australian Aborigine is one such group, initially subjugated by the British and subsequently by an expanding population of European residents. The Australian Aborigines have the oldest culture in the world. Due to the relatively recent occupation of their land and a developing and somewhat more enlightened approach by white Australia, significant remnants of their race and culture remain. Despite a typically anachronistic beginning and the subsequent efforts of missionaries, the new Australian occupiers are now attempting to redress the atrocities of the past and offer recognition, reconciliation and coexistence. However, it could easily be argued that this tends to be a one-sided interaction rather than a genuine connection, not so much due to apathy on the part of the Aborigines but rather their unwillingness to divulge the deep meaning of their culture to whites, whom they see as superficial and untrustworthy, and also to a corresponding failure on the part of white Australia to understand them; the classic racial and cultural divide.

Aboriginal culture and rock art

The land knows
The land makes you who you are
Without land, you are just nobody

(Author unknown)

Full-blood tribal Aborigines are deeply rooted in the land, it is their essential core. We see them as being "of the land" but they see themselves as "belonging" to it, especially the place where they were born. Babies at birth may be buried in the soil, face only exposed, to establish a connection with their land, to let it flow into them, like baptism. They come from the land, and they return to it. The land is their identity, it is where they belong. Their country is their reality, and they communicate with it. They have a deep and enduring spirituality and mysticism which is largely unrecorded and certainly not numbered amongst the greats, but perhaps it ought to be. Where we see a physical landscape; they see a spiritual one. Their Dreamtime is the Spirit of the Land, and to them, the land is everything, it is their "Book of Genesis". Their land is all around them, and their response always embraces the "significance" of the place. They are very much in tune with their biophysical world, and everything in it has a significance that dates back to the Dreamtime. To them the Dreamtime with its sacred spirits and mythologies represents not only their past history but also the ongoing spiritual life of all Aborigines in the present. They see the land as one total integrated system, with no lines drawn between past and present, plants and animals, or the living and the dead; and their rock art depicts and epitomizes this. The Aborigines have another attribute that we seem to have less of; they have family, a deeply felt love and spiritual identification with their immediate and recognized relations, not overtly expressed but solidly there in perpetuity. If one has an Aboriginal for a friend one has a friend for life, unconditionally.

There are many well preserved Aboriginal paintings in Australia, a large number of them located well away from popular tourist spots such as Kakadu National Park in Arnhem Land. Most of them could not be located without a guide. Typically, these paintings are protected under broad rock ledges and in shallow caves, some are still complete with grinding stones and original stone and bone tools. Some of them are known to go back 46,000 years and absolutely pre-date the Pyramids, which are only one tenth as old. Aboriginal rock art is not so much "art" or painting, as a graphical representation of a story from the Dreamtime relating to the fundamental icon, the great Rainbow Serpent. There are "stories" inside each piece. Commonly there are layers upon layers of depictions, writing and re-writing the history and progress of communities. The old, over-painted ones are never erased, they are still there in the background, just as Western history is not only recorded as being in the past but is continually occurring and being added to.

The Aborigines believe in an after-life and it is very real to them, but to them it is a spirit world with possible reincarnation, not necessarily in human form but maybe in the form of different animals, plants or even rocks. They have very stringent rituals and taboos concerning the welfare or possible malevolence of the spirits of the newly departed. Their concept of "conscience" or "soul" is what they call their "shade".

Full-blood Aborigines rely far more upon the capacities of the mind than we do and of course they have mental telepathy as well, and it is not just long range, they understand each other wordlessly. Who are we to call them primitive? To the Aborigines the Dreamtime is their Law, it tells them how to live, and their Law is far more rigorous than ours. Their rock paintings are their "Bible", each painting tells a story. The land is their university; their mind is their library, and their memory is unbelievably sharp; knowledge is their currency; dance is their celebration; and language is their culture, such a rich culture, it passes everything on down through the generations, and when a language is lost a culture is lost. Aboriginal languages are akin to a series of "word pictures" each having a meaning, rather than a series of words combined in a sequence to produce a sensible sentence.

Corroborees are a feature of Aboriginal life, and in full-blood tribal communities they can be heard every night. Some are recreational, for young and old, male and female, and everybody can participate. Others are more significant and for initiated men only. All of them are quite distinct from the jangling, clamorous song and dance of Western society, where the emphasis is on beautiful sexy young women. If you could experience and be really close to a genuine, heavy tribal corroboree with all its import and portent and full accompaniment of didgeridoos, boomerangs and clacking sticks, the only monotone musical instruments I know of, with all its rhythm, full-throated singing and chanting, deeply resonant, utterly riveting, wholly pervading all the night, so complex, forceful, authoritative, rigorous, and meaningful, and I would add a kind of subtlety, you might get an inkling.

The Aborigines can "sing" other Aborigines from a distance to make them do something, usually antagonistic, such as when one of them is trying to catch up with another with ill intent. It is known to work, probably by telepathy, as distinct from pointing the bone, which the victim has to be subtly made aware of and then fatalistically feels an outcome to be inevitable. They may also "sing" their prey, softly and out loud when they want to catch something, e.g. a fish. Do they somehow communicate with the fish? These people know things that we will never understand.

Since European invasion many tribes have lost all their old people and with them their memories, their language and their culture, never to be regained, replaced by the white man's way of life with its elements of junk food, disease, depression, hopelessness, religion, alcohol and drugs. The white man has almost completely pulled the rug out from under them. Being jailed for a misdemeanour they have never recognised and being removed and isolated from their land and their relatives must be the last straw. To "bridge the gap" by providing improvements in housing, employment, medical assistance and political representation, whilst clearly advantageous, on its own it is simply assimilation, turning Aborigines into whites.

Ethnological and sociological research is difficult at the best of times, but the Aborigines are a very protective and wary people who would happily offer misinformation rather than reveal their heritage. The information that researchers seek is not freely available, and many investigations so far have been somewhat superficial and condescending. There is hope on the horizon. Due to an almost lifetime of dedication and commitment some white researchers have gained the confidence of some older Aborigines, to the extent that some Aboriginal languages and aspects of culture are now being recorded, and even taught in schools alongside the ways of the English, about 20 languages out of an original total of about 350. Over 100 more are slowly being resurrected. Aboriginal languages were never meant to be written; their nearest approximation would be the message stick. Information, as education, was passed on by word of mouth and demonstration. Unfortunately, Aboriginal people are often being taught to say The Lord's Prayer and Grace before meals which they never had words for before, which has resulted in adoption of a form of pidgin between their language and English, not the language of their own spiritual beliefs, customs, stories, taboos and the deep meaning inherent in their sacred ceremonies and ancient rock art.

In understanding these people, the importance of body language such as inclination of parts of the body, eyes, and especially the lips is being left out. Their eyes might be downcast, furtive, exultant, angry, expectant; lips would move deftly left or right or up or down directing your attention; a nodding of the head, pushing out of the chin, an expression; the whole face could register a litany; enough emotions, instructions or information to tell a story. Their gender-specific rituals and ceremonies such as initiation and the passing on of culture are also ignored. Sacred sites are being recognized and protected, but that has more of a political flavour than a cultural one. It is a long and tortuous procedure with such a people to undo the cultural destruction and misinformation perpetrated by the white missionaries with their dogmatic and pious beliefs – they were not reluctant to reveal their gospel, quite the opposite, they forced it upon others, suppressed everything Aboriginal, dressed them in concealing clothes, banned their language and ceremonies, and ignored the horrendous ravages of the early settlers who regarded the Aborigines as vermin. What we now know about Aboriginal culture may be just the tip of the iceberg. The problem of the dispossessed, the refugees, remains.

Aboriginal "ecology"

In the south of Australia, we recognize four seasons. These are based upon the position of the earth in its elliptical orbit around the sun, becoming closer to or further away from the sun as it moves from its perigee of orbit to its apogee, and is consequently hotter or colder in a regular sequence. We clearly subdivide each season or segment of the orbit into particular months. Our specified months never change, every year is always divided into these same four seasons, and we have pre-written calendars to denote it. We even have an "Official start" date for the bush fire or storm seasons. In northern Australia we recognize two seasons – the "wet" and the "dry", but these are less specifically defined and are seen more in relation to the position of the North West Monsoon. The months in which they are likely to occur are recognized, but there is a lot of variation. The main "wet" is the equivalent of the southern summer because it is hot, but it is also very wet and humid, and the rest are all seen as one – the "dry".

But in northern Australia the Aborigines don't rely upon our formal seasons because their activities are directly related to natural processes that occur within the land. They know that there is an annual recurrence of these phenomena but do not invoke celestial processes to explain or schedule them. They recognize a number of subtle but quite evident changes in the land during both the wet and the dry seasons. They utilize readily observable biophysical "indicators" to tell them what is actually happening in the land at any one time. This pragmatic approach uses these indicators to lead them directly to sources of food and materials, not in relation to any pre-determined time frame or historical record, but when and where they actually occur. What do these indicate, and how do the Aborigines use them? I never knew what many of these indicators were actually indicating but I did come to appreciate several of them, and I considered them to be quite valid.

With the approach of the rains they note the turning of cumulus clouds to nimbus, and the first storms of the coming wet season tracking along the rivers from inland up towards the coast. Towards the end of the northern "wet" one wakes up one morning to discover that almost overnight the rain and showers have gone, and been replaced by clear blue skies with swarms of dragonflies, little green bulb-eyed spiders in the grass, black kites, and lots of mud-larks (Magpie larks) and wood swallows which were not there before, and the wind has turned to become south-easterly. We can well imagine what this indicates to the Aborigines in their quest for food and materials.

Of course, the Aborigines are aware of the tides, but there is another less obvious effect of the lunar cycle that they do not know the technicalities of but use to advantage. That concerns springs. At full moon when the moon's gravitational pull is at its greatest and we have the highest tides, the phreatic water table in the land rises in consonance with the tides, but with less short-term variability. Then the springs flow strongly with cool, clear, fresh water, but when we have neap tides the groundwater recedes, and these springs slow to a trickle or perhaps stop altogether. The Aborigines use this to advantage and arrange their travel patterns to coincide, according to water supplies they know will be there. At the next full moon they move on to the next spring, aided by moonlight, and hence around the country. Maybe our early explorers could have profited from this sort of information.

Aborigines also purposely "manage" the land, their main tool being the ecological one of fire. In the Northern Territory and the north of Western Australia where spinifex is the main grass, they light fires for two main reasons. One is to (as they put it) regenerate and refresh the land and to stimulate regrowth and the germination of seeds, so that the country can "grow back" and provide a new "crop" of bush tucker, both plant and animal. They also try to minimize the occurrence of uncontrolled wildfires that occur when a fire gets too large to be managed, often when one is started by lightning. They burn every year but only small separated patches at a time so that over the years all of the land is burnt and fuel levels are kept low, reducing the risk of damaging wildfires. The mosaic of burns from several seasons past can easily be seen on the ground and on aerial photographs, even on satellite imagery. They know where past burnt patches are and use them as "firebreaks" for new burns. They do not want their fires to get out of control and become damaging "big burns" so they may back-burn to stop a fire they have lit or set the fire towards a sand ridge or creek flat which would stop it naturally. These burns are always conducted in the cool season when the fires would be of low intensity and undamaging. This is systematic controlled burning, very similar to the controlled burns that fire authorities undertake in the south of Australia. Like European pastoralists, they recognize the spread of introduced grasses and weeds and their high burning heat, and regard them as unwelcome intruders on their land.

Historically, the advent of European man has had ecological consequences for the Aborigines. There is evidence that the introduction of grazing animals led to considerable ecological change, especially in the native grasses and herbaceous species that had provided much of their food. The introduced sheep and cattle found these species attractive, so that they were soon substantially eaten out and replaced by less useful pioneer species, in a retrograde ecological succession. This was exacerbated by the aggressive exclusion of Aborigines from these prime agricultural areas.

Modern thinking in the pastoral areas of southern Australia and around the world involves the use of "cell" or rotational grazing where small paddocks are intensively grazed for short periods, maybe for only a day or two at a time, then unstocked and allowed to regenerate until they are ready for grazing again. In this way the pasture never becomes old and senescent, but is always maintained at its maximum root mass and vegetative productivity. All plants are green and leafy and attractive to the animals and are eaten, and the most beneficial plants are able to compete favourably with the less desirable ones. Paradoxically, because very high actual stocking rates are used, the idea is to prevent overgrazing. If animals are left in a paddock for a prolonged period "to eat the weeds and all" the most desirable plant species are eaten out almost completely, their root mass becomes very low, and the pasture evolves into a sward of the least desirable species which of course are the survivors, the ones the animals leave until last; this is truly "overgrazing".

The Aboriginal equivalent is regular fires, which simulate the recurrent effects of the pastoralists' cell grazing of their animals. Although a different tool is used, fire instead of grazing pressure, the Aborigines established this management technique long before the white man did.

Given a whole list of environmental indicators and land management practices such as these, one can only conclude that the Aborigines are extremely observant and very clever ecologists. And it's not only the Australian Aborigines; capacities and attributes such as these were possessed by native peoples all over the world but have been largely over-ridden by aggressive European practices, and many are lost forever. The American Red Indian and the Kalahari Bushman are but two others; there are many. Humanity has squandered a great deal of its precious ethnic and cultural diversity.

Conclusion

It has been said that the major problems the world is facing today are in cyberspace, artificial intelligence and privacy. However, that assessment is temporal and superficial, and relates only to the more advanced countries of the world. The primary underlying causes of the predicament of mankind are much more deeply seated.

The mega-problems of the world are global, both in their generation and their impact. What one country does we all experience, it affects us all, indiscriminately. The atmosphere is global, so are the seas, and so are the problems.

The overriding problem mankind now faces is the widespread destruction of terrestrial, marine, and atmospheric systems and resources by a massive plague of humans.

Clearly, man is unable to control his population growth, or his penchant for war, environmental destruction and pollution. Just as clearly, nothing short of absolute disaster will have any effect upon his impact and the worsening state of the world.

Against the seeming inevitability that man will destroy his home on earth and possibly himself, we might envisage what we will have lost, what our "otherwise" future might have been. Large parts of the civilized Western world have created wondrous beauty by the exercise of intellect, imagination and concept, with magnificent human accomplishments in all forms of the arts, science and technology, medicine, architecture and urban precincts, and inspirational scholastic endeavour. Exalted man has walked upon the moon, and taken exploration and even occupation of space to credible prospective levels.

The obvious deliberation is whether all that creativity and beauty is to be obliterated by the inherent biogenetic propensities of the human species, which have culminated in the advent of the Anthropocene.

Unfortunately, the intellect of man appears to be at variance with his genetic imperatives. The ideal of a happy, healthy and liberated life in an unspoiled environment is in us all, but it is now a wish list, utopic, and mere fantasy. The elephant in the room is mankind's inherent compulsion to reproduce. The human genome is that of an animal, and his mind, which is his heart and soul, is unable to compete. The imperative of biogenetics will triumph over any amount of wisdom, whether we like it or not.

If Man can pull himself out of this predicament, he will have accomplished a truly superhuman feat.